Orbital motorways

Proceedings of the conference organized
by the Institution of Civil Engineers and
held in Stratford-upon-Avon on
24-26 April 1990

 Thomas Telford, London

Conference organized by the Institution of Civil Engineers and co-sponsored by the Royal Institute of Chartered Surveyors, the Société des Ingénieurs et Scientifiques de France, and the Transportation Research Board

Organizing Committee: A. Brant (Chairman; deceased), D. Bayliss (Acting Chairman), S. Mustow, M. Simmons, A. Whitfield

British Library Cataloguing in Publication Data
Orbital motorways
  1. Roads
  388.1
ISBN 0-7277-1591-7

First published 1990

Published for the Institution of Civil Engineers by Thomas Telford Ltd, Telford House, 1 Heron Quay, London E14 9XF.

Printed and bound in Great Britain by Mackays of Chatham

Orbital motorways

Contents

Keynote address

PROFESSOR P. HALL, MA, PhD, Institute of Urban and Regional
Development, University of California, Berkeley

How did we get to here? How did orbital motorways
come about? Who first built one? Who, for that
matter, first built a motorway? And why? These are
literally academic questions. The answers turn out
to be very often elusive and obscure. In this short
presentation, I would like to share with you my
attempts to piece together the historical record --
and in the process, to ask the key question: what
exactly has been the justification for ringing the
great cities of the world with motorways? For this,
too, proves to be a more elusive question than at
first you might think.

The Beginnings: Ring Boulevards and Parkways

An orbital motorway could not exist until motorways
were born -- on which topic, as said, there is
academic debate. But planners have been proposing
and even building orbital highways almost since town
planning began. Haussmann's mid-19th century
reconstruction of Paris is replete with them, but
even it builds upon earlier foundations. In the
mid-1850s Vienna built its Ringstrasse on the line
of the old fortifications, triggering a wave of
imitations in Central European cities. These
improvements had various justifications, but above
all they served to give the city a new image of
civic order and efficiency and monumentality: a
good place to live, a good place to do business in.

By 1905 Daniel Burnham, the American planner who
virtually created the City Beautiful movement, was
quoting these European examples in his abortive plan
for San Francisco: "A study of the cities of the
Old World develops the fact that the finest examples
-- Paris., Berlin, Vienna, Moscow and London --
consist of a number of concentric rings separated by
boulevards" (Burnham 1905, 39). On these examples,
Burnham based his plan for "the proposed system of
circulation for a larger and greater San Francisco".

Orbital motorways. Thomas Telford Ltd, London, 1990.

1

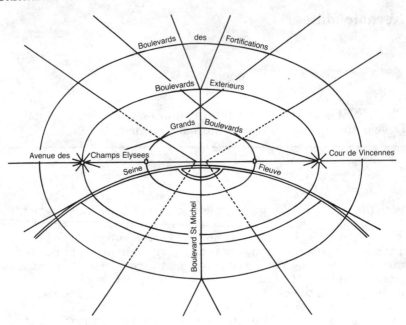

Fig. 1. 'Plan of Paris', 1905

(Fig. 1). It included an "Outer Boulevard" running
along the waterfront on the north, west, and east
sides (Burnham 1905, 39, 54-5).

But, in pursuit of the City Beautiful, Burnham did
not only appeal to European examples; he was able to
draw on an indigenous American tradition, too.
Starting with Frederick Law Olmsted's scheme for
carriage roads in New York's Central Park in the
1850s, came the notion of the parkway: "a road
running through the middle of a park" (Zapatka 1937,
97). The earliest examples, starting with Olmsted
and Vaux's Eastern Park Way in Brooklyn in 1868,
continuing with the park and parkway system for
Buffalo and Boston (Zapatka 1987, 103, 125), and
culminating in George E. Kessler's grand scheme for
Kansas City in 1893 (Rose 1979, 6), linked parks
with other parks, or residential neighborhoods with
parks (Zapatka 1987, 97).

But, just after the turn of the century, came a new
phase. The New York Legislature created the Bronx
River Commission in 1907 and planned a four-lane
parkway, first suggested in 1895, which was designed
by consulting landscape architect Herman Merkel and
completed in 1925 (U.S. Department of Transportation
1976, 132; Zapatka 1987, 113). Thence, during the

1920s, as Robert Moses extended the system around
New York City, the parkway became more and more a
long-distance commuter route. The Bronx, its
30-mile extension, and its successors -- the
Hutchinson River [1928] and Saw Mill River Parkways
[1930] -- were limited-access, grade-separated
routes forming a virtual network of parkways,
finally extended into Manhattan by the Henry Hudson
Parkway of 1936 (Zapatka 1987, 113-5). During the
1930s other areas followed the same principle:
Connecticut built the Merritt Parkway in 1934-8,
Southern California the Arroyo Seco Parkway begun in
1938 (U.S. Department of Transportation 1976, 397).
They are among the earliest pieces of road that
could fairly be described as motorways.

They were however all radial routes, connecting
suburbs and exurbs with downtowns. They awaited the
logical outcome: an orbital connector. It came in
the highway system proposed in the New York Regional
Plan of 1929, prepared under the direction of the
British emigre planner Thomas Adams: the
"Metropolitan Loop or belt line highway, which
circles the most intensively developed parts of the
Region" (New York Regional Plan 1929, 214).
Following the line of argument developed by Burnham
in San Francisco twenty years earlier, Adams argued
that this followed such classical plans as
L'Enfant's for Washington or Wren's for London (New
York Regional Plan 1929, 214).

The route, the plan said, "surrounds most of the
intensively developed residential areas, but in the
boroughs of The Bronx and Queens it has to pass
through existing gaps in areas that are already
built upon. On the average its course lies about 12
miles distant from City Hall" (New York Regional
Plan 1929, 221). "Connecting the boroughs of The
Bronx, Queens, Brooklyn and Richmond in New York
City the loop would greatly facilitate
intercommunication between these boroughs and
improve their connections with New Jersey centers.
At the same time it would relieve the pressure upon
the street system of Manhattan" (New York Regional
Plan 1929, 221). It includes the "Hudson River
Bridge" begun in 1927 (New York Regional Plan 1929,
224). (Fig. 2).

This then can fairly claim to be the first true
orbital motorway in the world. Its justification
was first to relieve traffic congestion in
Manhattan, and secondly to do so by decentralizing
and recentralizing activities in the outer part of

3

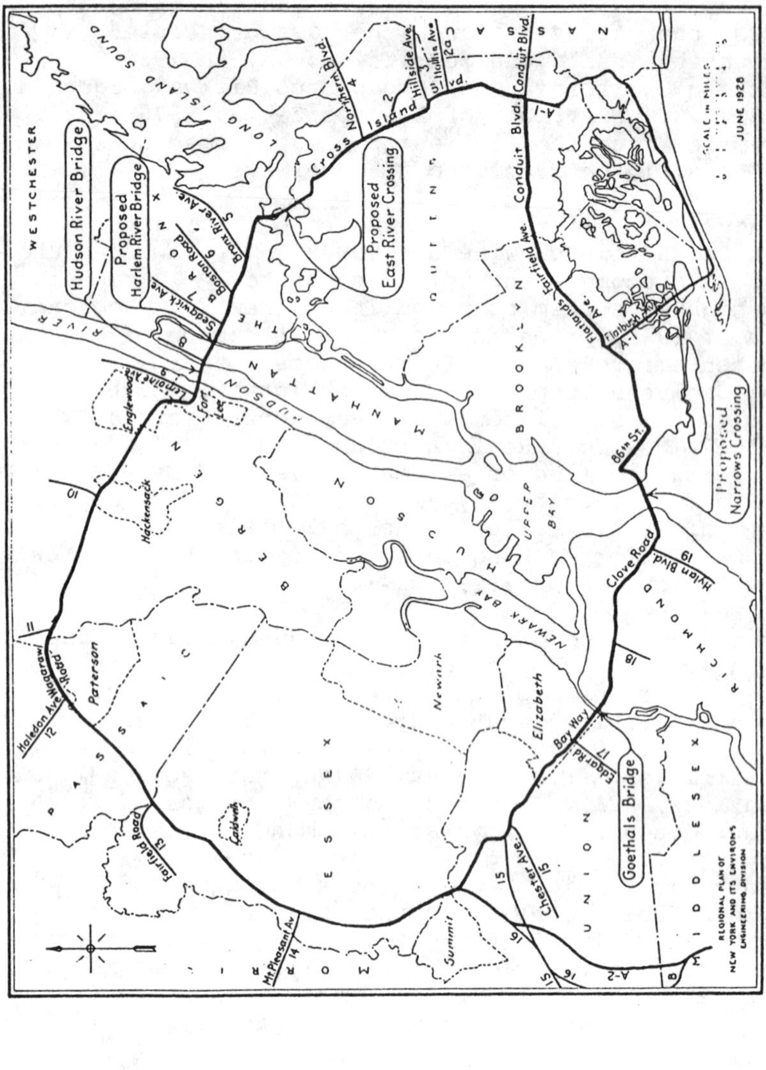

Fig. 2.   Proposed Metropolitan Loop Highway, New York, 1928

the region -- an aim which won the Plan the undying
opposition of Lewis Mumford. The year after the
plan's publication, Robert Moses -- who, as Robert
Fishman has shown in a new study, borrowed most of
his basic ideas from other people without
acknowledgement -- announced a plan for a
"circumferential parkway" encircling New York
(Zapatka 1987, 99). At a ceremony in 1938 marking
eight years of progress, with 33 miles open, he
explained that the plan "does not call for just an
automobile roadway, but a narrow shoestring park
running around the entire city and including all
sorts of recreation facilities, opening territories
which have been dead, relieving pressure on other
parts of the city, connecting the city with the
suburbs and the rest of the country, raising tax
values, encouraging building and spreading the
population" (Zapatka 1987, 117). That was precisely
the aim of Adams' team in 1929. Some parts of
Moses' network were not parkways at all, but express
routes running through densely populated
neighborhoods with no park-like features (Zapatka
1987, 117).

A map, Metropolitan Highway Loop, Progress in its
Development 1928-1941, published by the Regional
Plan Association Inc., New York (Lewis 1949, 131),
showed that "Excluding the parallel parkways,
progress on the main Loop to date shows 48 miles, or
41 per cent, of its total length constructed and an
additional 15 miles, or 13 per cent, having received
official preliminary action. Of this total, seven
miles have been advanced in the past four years"
(Lewis 1949, q. 131-2). (Fig. 3). The highway,
over 100 miles long, was completed by 1944 (Zapatka
1987, 117).

There were very faint echoes of the parkway
movement in Britain too. The landscape architect
T.H. Mawson built a parkway around Stanley Park in
Blackpool in the early 1920s, and Louis de Soissons
built parkway approaches into Welwyn Garden City; at
the end of that decade Barry Parker, who knew
America well, incorporated parkways into his plan
for the satellite town of Wythenshawe outside
Manchester. None of these was an orbital parkway.
But the Bressey-Lutyens highway plan of 1937, which
incorporated pieces of Raymond Unwin's abortive
Greater London Regional Plan of 1929-33, did borrow
from it the notion of an orbital parkway through the
Green Belt Ring which the London County Council was
then proposing, and which it began to implement in
the Green Belt Act of 1938. Parts of this orbital

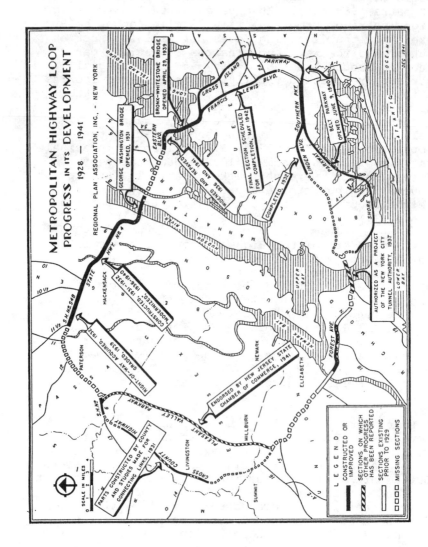

Fig. 3

parkway can still be seen between Denham and
Rickmansworth and between Watford and London Colney;
it was however designed as an ordinary highway with
intersections at-grade.

The European Beginnings of the Motorway

Meanwhile, across the Channel, they were building
motorways in earnest. There is a great deal of
scholarly dispute as to exactly where the motorway
began. The Americans, as already seen, argue that
Robert Moses' New York parkways were among the first
to incorporate controlled access and grade
separation, albeit partially. The Italians, on the
other hand, have long claimed that the first
motorway in the world was the one opened between
Milan and Varese on the route to the Italian lakes,
by King Victor Emmanuelle III on 21 September 1924.
The brainchild of Pero Puricelli, it was proposed in
a document of 11 March 1922 issued by the Touring
Club Italiano (Bolis 1939, 574). By 1925 some 80 km
were open; branch routes to Como and Maggiore
followed, extending the system to 130 km.

By the late 1930s a whole series of autostrade were
completed, nearly all in the northern plain and
forming a continuous autostrada connecting Torino
with Milano, Bergamo, and Brescia with branches to
the lakes, plus isolated stretches from Naples to
Pompeii, Florence to the coast, Venice to Padua, and
Genoa to Serravalle (Bolis 1939, 577, 580-1;
Charlesworth 1984, 11). The main trunk line
by-passed Milano to the north (Bolis 1939, 580-1)
but nowhere on this system was any hint of a ring
motorway; that had to await the post-World war II
period, with construction of the Rome ring on the
Autostrada del Sole. They were single carriageways
but completely segregated with at least three lanes
and with controlled access and no at-grade crossings
and a design speed of 100 k/hr (Charlesworth 1984,
11).

In other words, these Autostrade were not full
motorways in our sense of the term. Perhaps that
distinction belongs to the Avus in Berlin, a
five-mile special road used partly for motor racing
and partly for tourist traffic, built through the
Grunewald in Berlin between 1919 and 1921. Then, in
1926, came the formation of the Hafraba, a private
organization which planned a 500-mile commercial
toll highway connecting Hamburg, Frankfurt, and
Basel, which failed for lack of funds (GB Admiralty
1945, 468). Undeterred, the organization succeeded
in building a short length of motorway between

Cologne and Bonn, and announced grandiose plans for a 3,000-3,500 mile national network (Hass-Klau 1989, 117).

The Nazis capitalized on these plans soon after coming to power (GB Admiralty 1945, 468). Legislation was introduced on 27 June 1933, and in August the same year the Reichsautobahnen Gesellschaft was established as a subsidiary of the Deutsche Reichsbahn with capital of RM 50 million and freedom from taxes. Dr Todt, who organized the scheme, was made Inspector-General of German Roads in 1934, and was final authority (GB Admiralty 1945, 468). He was placed directly under Hitler with absolute power (Cement and Concrete Association 1936, 12). The German Railway was made the holding company because of its financial and organizational capacity (Cement and Concrete Association 1936, 12). The scheme was operated by railway engineers and the Autobahnen incorporate characteristic rail features (Brodrick 1938, 201; GB Admiralty 1945, 472).

Surfacing began in July 1934 between Frankfurt and Darmstadt, and this stretch -- part of the Hafraba -- was the first finished in May 1935 (GB Admiralty 1945, 449, 469). The plan proceeded with remarkable speed with a target of 4,300 miles and a schedule of 650 completed miles a year; some 670 miles were already open in September 1936, 950 miles a year later, and 1,120 miles under construction (Cement and Concrete Association 1936, 3, 18; Cement and Concrete Association 1937, 7; Brodrick 1938, 204). By 1943, 2,380 miles were complete (GB Road Research Laboratory 1948, 3), including the main lines Stettin-Berlin S Ring-Ruhr-Frankfurt -Munich-Salzburg as well as Berlin-Leipzig-Munich (GB Road Research Laboratory 1948, 3).

The justification for this enormous program certainly could not be found in traffic congestion. There was one car to every 26 people in the country, one to 21 in the cities (Germany, Generalinspektor 1938a, 12). There were nearly four times as many cars per head in Britain as in Germany (Brodrick 1938, 210). The county surveyor for Ayrshire noted that: "the motor-roads of Germany are far in excess of present requirements, or indeed the requirements of many years to come" (Brodrick 1938, 210). Traffic was extremely light when a British Parliamentary delegation visited: "One passed over the autobahnen for miles", wrote the Member for Perth, "without meeting another vehicle" (Brodrick 1938, 210). The opening of the Cologne-Düsseldorf section showed the"great carrying capacity: an average of 23 cars per minute" (Brodrick 1938,209).

German authorities at the time repeatedly stated that the roads were built for peaceful uses; this was often questioned abroad (Brodrick 1938, 209-10), but there is no real evidence to show whether military uses were considered or not; if indeed the roads were built with war in mind, it seems difficult to understand why some links were left unfinished and some almost unused during the war (GB Road Research Laboratory 1948, 3). More truly, it seems that the entire plan was really an enormous programme of unemployment relief with monumental overtones.

Partly for that last reason, the basic plan -- a grid with two north-south and four east-west roads -- also contained some monumental motorway rings. Most cities had half or quarter circles which were sufficient to ensure easy connection between the urban area and the Autobahnen (GB Admiralty 1945, 472). But, Nazi propaganda stressed, unique solutions were required in the so-called World Cities -- Berlin, Hamburg, Vienna -- and also in Munich, capital of the Nazi movement and center of German culture (Germany, Generalinspektor 1938a, 12). For these cities Hitler had ordered a new monumental city form, to which the Autobahnen must fit (Germany, Generalinspektor 1938a, 12).

Around Berlin, three Autobahnen -- Hamburg-Breslau, Stettin-Munich, and Frankfurt an der Oder-Ruhr -- were to be linked by a ring some 180 km., more than 100 miles, long; more than 20 interchanges would serve the city (Germany, Generalinspektor 1938a, 12). The distance of the ring from the city was determined by the many lakes in the region, the position of Potsdam, and the need to provide for extensions of the city (Germany, Generalinspektor 1938a, 12). The radials intersecting the ring were to be connected to the two new axes of the city which Albert Speer, Hitler's chief architect-planner, had designed (Germany, Generalinspektor 1938a, 13). These radial links were designed for easy access but were not proper motorways, but had periodic intersections at grade; the exception was the Avus, the oldest motorway in the world (Germany, Generalinspektor 1938a, 13).

By August 1936, the southern and eastern parts of the Berlin ring were complete (Cement and Concrete Association 1936, 2; Germany, Generalinspektor

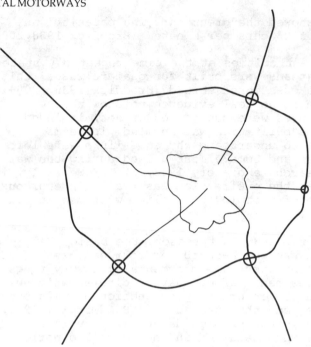

Fig. 4. Connection of Munich to the Imperial Motorway, 1938

1938a, 31). By 1943 the Berlin ring was
approximately two-thirds constructed (GB Road
Research Laboratory 1948, 3).

Munich also required a ring, partly because of the
extension of the Reich into Austria, which meant
that it was on a through route (Germany,
Generalinspektor 1938a, 13). It would be only some
60 km., 40 miles, in length, reflecting the
different geography (Germany, Generalinspektor
1938a, 13). (Fig. 4). The ring allowed fast car
traffic to be kept out of the inner city, where
traffic needed to be calmed [that concept was
actually being used by German engineers in 1938],
and also allowed through goods traffic to be kept
out of the inner city (Germany, Generalinspektor
1938a, 13).

Vienna presented special difficulties, not least
the nearness of the Austrian frontier in the Vienna
woods; the solution was not yet ready in 1938, and
indeed was never found (Germany, Generalinspektor
1938a, 14).

In Hamburg the solution centered on the Elbe High

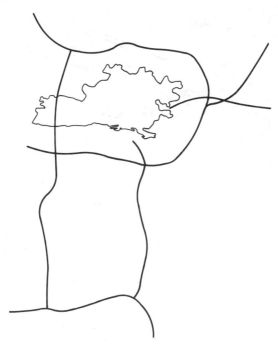

Fig. 5. Connection of Hamburg to the Imperial Motorway, 1938

Bridge (Germany, Generalinspektor 1938a, 14); here
too a complete ring was planned, forming two loops
(Germany, Generalinspektor 1938a, 14). (Fig. 5).
The bridge had a middle span of 750 m. and was 70 m.
clear above the water, allowing passage of the
highest ships (Germany, Generalinspektor 1938a, 16).

Thus the justification for the great German
orbitals of the 1930s is obscure. It was ostensibly
monumental and symbolic, but may have been for
military logistics. In any event, it was not
justified by any concept of traffic congestion.
After World War II, the Munich and Vienna rings were
never completed. The Berlin ring was -- and for the
time being is -- totally within East Germany; parts
performed an important function in carrying transit
traffic from West Germany to West Berlin, but the
northern and eastern sides of the ring were
apparently little used.

America: The Coming of the Interstates

Back in America, the General Motors Futurama
exhibit was the major event at the 1939 New York
World's Fair. It promised that motorists would be
able to drive at 50, 75, and even 100 miles an hour

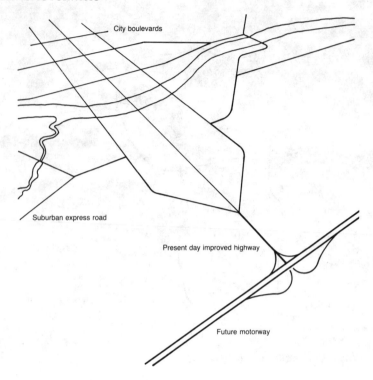

Fig. 6. Vision of motorway feeder to city, 1935

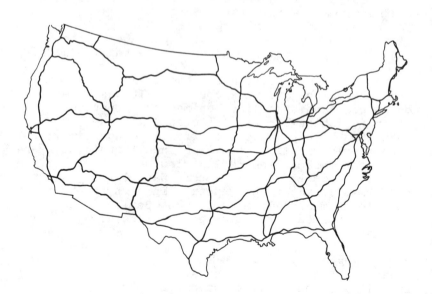

Fig. 7. Proposed interregional highway system, 1939

on a completed segregated system of motorways; these would penetrate cities, "so routed as to displace outmoded business sections and undesirable slum areas" (Rose 1979, 1). (Fig. 6). Ironically, Norman Bel Geddes, its visionary designer, though an enthusiastic proponent of superhighways for interregional travel, took a very European view: he was against their use in cities (Jones 1969, 205).

At almost exactly the same time, Los Angeles published its celebrated freeway plan. In its original form, it owed much to the long parkway tradition (Jones 1969, 46). That same fateful year, 1939, a crucial report from the Bureau of Public Roads started to pave the way to the Interstate program. It was Toll Roads and Free Roads, based on the first analysis of the nationwide pattern of highway usage, developed from surveys by the Bureau and State highway authorities (U.S. Congress 1939, IX). It defined three east-west and three north-south "superhighways" totalling 14,336 miles and estimated to cost $2.9 billion (U.S. Congress 1939, IX), and showed that it was not possible to finance them fully through direct-toll collection (U.S. Congress 1939, IX). Tolls were impracticable because much traffic was either short-distance or unwilling/unable to pay (U.S. Congress 1939, 89). The report went on to detail a 26,700-mile system of toll-free highway development, "believed to include substantially every major line of interregional travel in the country" and joining all major cities (U.S. Congress 1939, 108, and Fig. 57, facing 108). (Fig. 7).

"One of the striking characteristics common to all highway traffic maps", the report noted, is the sharp enlargement of the bands representing the volume of traffic on the important highways as they approach the larger cities", unaccompanied by an enlargement of the arteries themselves; hence congestion (U.S. Congress 1939, 90). "The remedy commonly proposed for these conditions is the construction of a bypass highway... In rare cases this remedy alone may prove sufficiently effective, but, as hereafter elaborated, bypass routes are of advantage mainly to a relatively small part of the highway traffic normally approaching a city, i.e. to that small part of the traffic that is actually desirous of avoiding the city" (U.S. Congress 1939, 91). Analysis showed that on the highway between Washington, D.C. and Baltimore, at Washington the "bypassable maximum" was 2,269 out of 20,500 entering vehicles, 11 per cent; at Baltimore, 2,670

13

out of 18,900, 14 per cent (U.S. Congress 1939, 91).
"In the larger cities generally only a major
operation will suffice - nothing less than the
creation of a depressed or an elevated artery [the
former usually to be preferred] that will convey the
massed movement passing into, and through the heart
of the city, under or over the local cross streets
without interruption by their conflicting traffic"
(U.S. Congress 1939, 93).

The report did allow a concession: "Next to
provisions for the safer and more efficient conduct
of large traffic streams into and across cities", it
admitted, "the new facilities most urgently required
are belt-line distribution roads around the larger
cities and bypasses around many of the smaller
cities and towns" (U.S. Congress 1939, 95). But it
again stressed that "the traffic on a main highway
approaching a large city, that will use a bypass
route if offered, is a small part of the total"
(U.S. Congress 1939, 95). So the main function of
these "belt-line distribution roads" was to serve
not as by-passes but as distributors between radials
and between different sections of the city:

... for those parts of the traffic on each entering
highway that are (a) interchanged with other
entering highways not nearly opposite across the
city and (b) originated in or destined to sections
of the city similarly situated, the facility that
will generally provide the best service is a
circumferential or belt-line route forming an
approximate circle around the city at its outer
fringe (U.S. Congress 1939, 96).

In some cases, however, an alternative would be "to
construct the distributing belt line within the city
-- generally somewhere within the ring of decadent
property surrounding the central business area"
(U.S. Congress 1939, 97). A true belt-line route
"must be protected from the encroachment of
bordering developments that would quickly engulf it
and destroy its special character" (U.S. Congress
1939, 98); thus such roads "must be built as
limited-access highways, cut off from the bordering
land except at a very limited number of points, and
separated from all but a very limited number of the
cross streets and highways intersected by them"
(U.S. Congress 1939, 98). Plate 52 from the report
shows a tentative study of such a highway for
Baltimore (U.S. Congress 1939, facing 100). (Fig. 8).

In 1944 came a second milestone report: Inter-

14

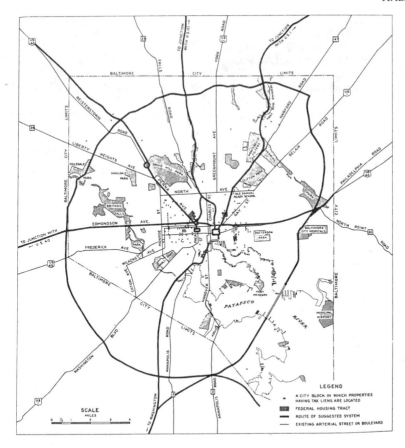

Fig. 8. Tentative study of orbital highway for Baltimore, 1939

regional Highways. In a bicentennial history, the United States Department of Transportation has said that: "Without doubt Interregional Highways was and remains the most significant document in the history of highways in the United States" (U.S. Department of Transportation 1976, 274). The Committee that produced it was headed by Thomas MacDonald and included Frederic Delano, Chair of the National Resources Planning Board, Harland Bartholemew, city planner of St. Louis, and Rexford Guy Tugwell, Chairman of the New York City Planning Commission (U.S. Congress 1944, IX). It started by detailing the neglect of interregional highway planning and the mistakes made in building the earliest urban highways. It argued that "[the rural] emphasis needs to be reversed and the larger expenditure devoted to city and metropolitan sections of arterial routes" (U.S. Congress 1944, 3).

It recommended a 33,920-mile system, of which 4,470 miles would be urban. It directly connected all cities of 300,000 and more, and was the smallest system that would achieve this (U.S. Congress 1944, 6). Without any aid from computers, it went into great geographical detail on the system's level of service in relation to cities of different size, rural population distribution, manufacturing centers and farm production, and car ownership (U.S. Congress 1944, 6-39; U.S. Department of Transportation 1976, 274).

In planning the urban component, it again noted -- as had its 1939 predecessor -- the very rapid increase in traffic densities as cities were approached (U.S. Congress 1944, 42-3). The resulting zones of dense traffic extended as wide as 35 miles for cities of three million and more, to 30 miles for cities of one to three million, and 25 miles for half to one million (U.S. Congress 1944, 43). It argued that transportation had been a major factor on urban growth, and that "It is highly important that this force be so applied as to promote a desirable urban development" (U.S. Congress 1944, 55). Through a study of Baltimore it demonstrated that the proportion of by-passable traffic for a city of this size was low, only 21 per cent (U.S. Congress 1944, 58-9). Another study of 27 cities of various population classes showed that the proportion was 7.2 per cent for cities of 300,000 to 500,000, and 4.2 per cent for cities between 500,000 and 1,000,000 (U.S. Congress 1944, 60). Much of the city-bound traffic would go to or close to the center: "It is reasonable to conclude, therefore, that the interregional routes, carrying a substantial part of this traffic, should penetrate within close proximity to the central business area" (U.S. Congress 1944, 61); the details would vary for each city (U.S. Congress 1944, 61). For the non-central traffic:

To serve this traffic bound to or from points other than the center of the city, there is need of routes which avoid the business center. Such routes should generally follow circumferential courses around the city, passing either through adjacent suburban areas or through the outer and less congested sections of the city proper (U.S. Congress 1944, 645).

Such routes could serve both through traffic and as distribution routes for intra-urban traffic (U.S. Congress 1944, 65). Later it illustrated a plan for a large city, with two circumferential routes, an inner and an outer, actually bisected by three

Fig. 9. Motorways proposed by Institution of Highway Engineers, 1936

through routes which made a triangular enclosure of the CBD (U.S. Congress 1944, 72).

The 1944 Federal-Aid Highway Act, which established the National System of Interstate Highways, followed almost exactly the recommendations of this committee (U.S. Department of Transportation 1976, 277). It also established a spending formula which reserved 25 per cent to urban areas; this survived until 1973 (U.S. Department of Transportation 1976, 277). It limited the system to 40,000 miles and provided that the routes should be selected by State highway departments jointly with adjacent states (U.S. Department of Transportation 1976, 156-8). A 37,000-mile system was circulated for consultation during 1946 and a 37,681-mile system, including a mere 2,319 miles of urban highway, was finally agreed and approved by the Federal Works Administrator on August 2, 1947 (U.S. Department of Transportation 1976, 158). As in Britain, the start of the system was long delayed; but finally, "In

1956, Congress voted $27.8 million to build the
National System of Interstate and Defense Highways,
the most ambitious public works program since the
Roman Empire" (Jones 1969, 18).

One report, published on the verge of the
Interstate Program's start, provides invaluable maps
of some 100 major metropolitan areas showing the
systems then planned. Relatively few show complete
or near-complete beltways: they include Little
Rock, San Francisco, Denver, Washington, Atlanta,
Chicago, Indianapolis, Davenport-Rock Island-Moline,
Topeka, Baton Rouge, Portland, Baltimore, Boston,
Minneapolis-St Paul, Kansas City, St Louis, Omaha,
Albany, Buffalo, New York, Columbus, Oklahoma City,
Sioux Falls, Memphis, Nashville, Dallas, Houston,
and San Antonio (U.S. Department of Transportation
1955, passim). In a number of cases it is an
arbitrary distinction as to whether the city has a
true "beltway" or just an extended bypass. In
general, just as the 1939 and 1944 reports had
urged, the system penetrates to the hearts of the
cities. Thirty-five years later, in 1980, a
comprehensive review found that: "To date, partial
or complete circumferential highways have been
constructed in 35-40 cities... Twenty-seven cities
have limited access beltways that are included in
the Interstate Highway System" (U.S. Department of
Transportation 1980b, 14).

Beginnings in Britain
The British as usual were a little late. Lord
Montague argued for elevated roads in London and for
a London-North West four-lane motorway as early as
1923-4 (Drake et al. 1969, 35). The latter idea was
killed by railway opposition (Drake et al. 1969,
35-6). In 1936 came a plan by Institution of
Highway Engineers for 2,800 miles on 51 lengths,
rejected by Transport Minister Hore-Belisha
(Charlesworth 1984, 13). (Fig. 9). In 1937 the AA
and RAC organized a trip to Germany, bringing
together MPs, County Surveyors, and Councillors, and
others; it unanimously concluded that some motorways
were needed (Charlesworth 1984, 14). The Minister,
Leslie Burgin, visited Germany the next year and
approved a Lancashire County Council plan for a
60-mile route, but no money was available to start
it (Charlesworth 1984, 14). In 1937 the County
Surveyors' Society accepted the principle of
motorways and in May 1938 it drew up a plan for
1,000 miles, which was close to the postwar plan but
differed in detail (Charlesworth 1984, 14-16).
(Fig. 10). The 1936 plan contained a London

**Fig. 10.   National Plan for Motorways,
County Surveyors' Society, 1938**

orbital; the 1938 plan did not (Charlesworth 1984,
maps 12 and 15).

In 1946, after the allied forces had won the war on
the Autobahnen, Britain at last officially espoused
motorways. The original roads programme was
announced by Transport Minister Alfred Barnes on 6
May 1946 and displayed in the House of Commons Tea
Room that year. It showed an hour-glass network
approximately 800 miles in length; and it included
the London orbital (Starkie 1982, 2; Charlesworth
1984, 23-4). (Fig. 11). The Special Roads Act
followed in 1949, the same year as James Drake
produced a comprehensive highways plan for
Lancashire including 94 miles of motorways including
parts of rings for Liverpool and Manchester
(Charlesworth 1984, 27). J.F.A. Baker was made head
of a special motorways team and visited Germany to
see the Autobahnen in 1950; he was the father of the
system (Charlesworth 1984, 24-5).

Fig. 11. Ministry of Transport Road Programme, 1946

In 1946 the government had intended expenditure to
start construction of the national network to start
five years later; but this fell victim to recurrent
fiscal and economic crisis (Starkie 1982, 3-4). Not
until 1955-57 did detailed planning and financial
commitment come; not until the late 1950s were the
first stretches of motorway opened (Drake et al.
1969, 46). Interestingly, one of the first was the
Stretford-Eccles bypass of 1960-61, the first urban
motorway in Britain and part of a planned ring for
Manchester (Drake et al. 1969, 205). When the
program resumed, "it had to do so by digging deeper
into its inheritance from 1946, namely the tea room
plan" (Starkie 1982, 5). The map of completed
motorways in December 1978 shows a strong
correlation with the 1946 plan, but with the
conspicuous exception of the London orbital, which
was barely begun (Starkie 1982, Fig. 13.4, 141).
Thus one could say that the bulk of the 1946 plan
was completed in thirty years, but the M25 took
exactly forty.

The years from the mid-1950s to the mid-1960s were the golden age of motorway planning, which included many important orbital elements. In the West Midlands proposals were drawn up in 1959 by the authorities and MOT, including the M42 as a partial ring; this was further developed in a 1964 study (Charlesworth 1984, 199). In Greater Manchester the SELNEC Highway Engineering Committee produced a plan which included a ring around the city (Charlesworth 1984, 201). Liverpool is also shown with a complete orbital (Drake et al. 1969, 174). This was programmed before MALTS, a study of the late 1960s. The first section opened in 1972 (Charlesworth 1984, 204).

The Manchester and Liverpool sagas demonstrate that the transportation studies of the 1960s tended to end up backing schemes that had already been committed: the SELNEC study of the late 1960s shows that the whole of the Manchester ring, save for a short section in the north-east, was previously announced (Starkie 1982, 69). Much of this network was in fact built by the late 1980s.

The 1970s brought a huge reaction against urban roadbuilding, especially in London (Starkie 1982, 73). The Layfield Inquiry into the Greater London Development Plan concluded in December 1972 that only Ringway One and the M25 were justified, but this was academic since in July 1972 the London Labour party took an anti-motorway stance (Starkie 1982, 84). One factor was the decision to go ahead with the M25 (Charlesworth 1984, 197). The M25 got "highest" priority in 1976, "first" priority in 1978, and "top" priority in 1980, when 36 miles were open and 21 miles were under construction (Charlesworth 1984, 198).

The Saga of the M25

Finally the M25, London's 117-mile orbital beltway, was opened in October 1986. The final 13-mile section, between the A41 at Watford and the A1 at South Mimms, closed the world's longest by-pass, connecting it with the Midlands and North of England via the M1. It took in all thirteen years to build -- the first short section, next to the last one, was started in 1973 and opened in 1975 -- and over forty years to plan.

Many people suppose it to be Patrick Abercrombie's invention: the cornerstone of his 1944 Greater London Plan. This is only partly true. A ring for London appeared in plans before Abercrombie's: the

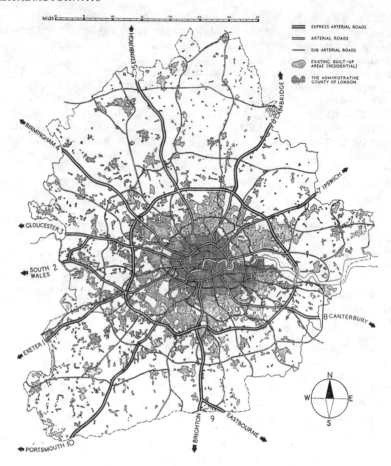

Fig. 12. Abercrombie's Greater London Plan, 1944

first, perhaps, in the report of the Royal
Commission on London Traffic in 1905. The first
sign of an orbital motorway is in the report of the
Greater London Regional Planning Committee 1933,
though as a "main road parkway" (Greater London
1970, Fig. 9 opp. 19). As already seen, the
Institution of Highway Engineers plan of 1936
included a London orbital. The Bressey Plan
[Greater London Highway Development Survey] of 1937
clearly shows North and South Orbital Roads, and
these are very close to the present M25 (Greater
London 1970, Fig. 12a opp. 22, and 23). The
Abercrombie Plan's D Ring was close to Bressey,
one-fifth of it consisting of Bressey proposals: it
was to provide shortest possible connections from
radial to radial, accommodate rapid transit between

airports, and give access to Thamesside industries and docks (Greater London 1970, 30). (Fig. 12).

The important detail is that Abercrombie's major orbital -- his so-called D Ring, fourth in his series of five -- was five miles closer to the center than the M25. The only one actually planned as a motorway, it would have passed through the fabric of outer London, separating Hayes from Southall, Harrow from Ruislip, Edgware from Watford, Ilford from Romford, Croydon from Coulsdon, Epsom from Watford. Only for a few miles in the north, between the A1 and the A10, does the M25 roughly follow the Abercrombie line. Pieces of the original D-Ring survive in different projects, one of which, the Hayes By-Pass in West London, is under construction and will open in the early 1990s.

The M25 is thus not Abercrombie's express-arterial D Ring, but his outermost sub-arterial E Ring, which -- following the model of Robert Moses in New York and Barry Parker in Wythenshawe -- he saw as a parkway. Though a dual carriageway, it would have been quite different from the M25: it would have had "ample provision for lay-bys to both carriageways to allow for parking of cars for short periods", while to "allow for picnic parking off the main road there should be easy access to the adjoining local roads". Thus, in the thirty years between the Abercrombie plan and the start of the M25, his two outer rings were telescoped; in effect, the parkway was submerged under the motorway.

This illustrates an important point: Abercrombie's own justification for this entire grand structure -- five rings and connecting radials -- was only partly the relief of traffic congestion. As Douglas Hart has so well shown, he was equally concerned to make the new highways a key element in the attempt to give London a structure -- or, more accurately, to reveal a latent organic structure (Hart 1976, passim). Within the built-up mass, they would separate and define London's historic neighborhoods: its villages. Outside it, they would define and delimit the entire built-up mass. The D Ring would play a particular role, because in effect it would define the outer edge of London.

In this regard, with benefit of hindsight, Abercrombie was probably wrong and the subsequent modifications were right. He argued that the M25 line was too far out "to afford relief to Inner London or to Regional traffic requirements" -- a prophecy vitiated by forty years of traffic growth that neither he nor his contemporaries ever imagined. Oddly, he claimed that his D-Ring "would girdle the general limits of the built-up area of London", though this was demonstrably not true even then. And, because of that fact, his route would probably have generated as much controversy as his inner arterial B-ring -- which, renamed the Motorway Box and then Ringway One -- was abandoned by the GLC in 1973. In fact, the likelihood is that it would have proved so controversial that it would never have been built.

So it is possible to argue that the M25 is the route that Abercrombie would have chosen if he had more time to ponder. It sits in the middle of the ten-mile-wide green belt of the 1944 plan, and even if its free-low interchanges occupy chunks of it, it has been finely landscaped and integrated into the surrounding countryside. What Abercrombie did not foresee was the regional planning context of the 1980s.

As has often been said, his was a static view of the world: once London's war-damaged and slum-ridden inner ring had been rebuilt, once the great planned movement to the new and expanded towns had taken place, the London region would settle down to decades of slow, almost glacial, change. Abercrombie seems never to have conceived of a London that would continue to shrink in population and in basic employment -- even though the Regional Planning Association of America, whose basic philosophy he shared, had predicted the decline of great cities already in the mid-1920s. Thus he could not see that his express orbital, wherever placed, would become a magnet for development, aiding the movement of activity from core to periphery.

Orbital Motorway Impacts

Orbital motorways, then, have been planned on very varied grounds: the Americans justified them on the basis of traffic flows, though their own analyses suggested that most traffic would not use them; Abercrombie and his predecessors proposed them on regional planning grounds, though they had few traffic surveys to justify their hunches; the German engineers built them for monumental, nationalistic, and perhaps military reasons. No one, apparently, had much concern for the impact of the new roads on the land use patterns themselves -- though some pioneers, like Benton McKaye, had suggested these impacts as early as the 1920s (Hall 1988, xxx).

Here in Britain, in 1981, Nathaniel Lichfield and Alfred Goldstein joined forces and tried to predict the impact of the future M25. They saw rapidly rising rental and capital value for all commercial property, accelerating after completion and most evident around and beyond the Green Belt. Offices in Central London, they thought, would find it increasingly attractive to move to Home Counties towns close to M25 with good labor supplies and environment. Warehousing and manufacturing would move to motorway connections; and the M25 would prove a magnet for superstores. The main problem for developers would be to get permissions, and this would drive them eastwards where planning authorities were more open (Lichfield and Goldstein 1981, 35).

In 1982 the Standing Committee on South East Regional Planning Committee returned to the same theme. They started from the hypothesis that the motorway would create "a major new structuring element in the region", significantly increasing development pressures in areas of strict restraint while making it difficult to regenerate inner London (SERPLAN 1982, 50). The area west of London, already attractive to high-tech-related development, would come under even greater pressure (SERPLAN 1982, 51). But big journey-time improvements would increase the competitive advantage of the relatively depressed eastern sector, and extend labor markets, thus widening firms' locational options. There would be opportunity for extensive industrial and distributive development especially in the lower Thames corridor. And in parallel, there would be increased possibility for offices and other intensive development in outer London centers -- Croydon, Sutton, Kingston, Wembley, and in certain OMA centers close to M25 intersections (SERPLAN 1982, 53). Therefore, SERPLAN argued, planning authorities should promote opportunities in the eastern sector, especially for extensive industrial and distributional uses (SERPLAN 1982, 54), coupled with transport measures to spread the impacts inwards, especially in the depressed eastern sector (SERPLAN 1982, 56). This, recall, was speculative -- though it has proved an extremely good predictor of what has actually happened in the three and a half years since the motorway was completed.

Even earlier, in 1977, the British Leitch Committee on Trunk Road Assessment had reviewed the argument that motorway construction induced industries to locate in those regions with improved communications. It found that "the case for the 'restructuring' effects of trunk road construction on economic growth in developed countries such as Britain is weak and at best not proven" (G.B. Advisory Committee 1977, 92). It looked at a considerable number of studies which had tested the hypothesis that, by increasing accessibility, trunk road developments in a region would have a significant impact on economic growth there (G.B. Advisory Committee 1977, 206). It concluded that "few of the studies would appear to support the hypothesis" (G.B. Advisory Committee 1977, 206). "At the national and regional levels, therefore", it concluded, "all the evidence points to the conclusion that improvements to the trunk road system can have only a limited effect on industrial location and growth" (G.B. Advisory Committee 1977, 207).

There was one qualification: any effects were likely to be local, for instance where hypermarkets took trade from town-center shopping centers (G.B. Advisory Committee 1977, 207). So the sole case in which motorways did appear to have some impact was orbital motorways around cities; and since this was a very local impact, dragging activities out of the cities and into the suburbs, it was not of direct interest to the committee.

Over in the United States, an extremely comprehensive impact study of beltway impacts was made shortly after the Leitch Committee reported. It was by Payne-Maxie Consultants for the U.S. Department of Transportation (Guiliano 1986, 266). It contained three phases: a review of all previous related research, a set of statistical analyses on a set of 54 U.S. cities [27 with beltways], and in-depth case studies of eight cities (Guiliano 1986, 266). The statistical analysis revealed that the land use impacts of beltways have been largely insignificant; growth effects as manifested by regional population and employment changes over 1960-77 were related to change in manufacturing employment and [negatively] to city age; the existence of a beltway, the beltway's relative location, and its length had no consistent effects on growth (Guiliano 1986, 266).

Beltways could have a "one-time effect" on central city office space, weakening the downtown market and drawing some employment out; in the case study cities, this effect was small and did not harm revitalization efforts (U.S. Department of Transportation 1980b, 115). Historically, central cities surrounded by suburban beltways posted lower gains in retail sales and employment than those with beltways within their jurisdiction or with no beltways; though the most important determinant of central-city retail sales was changes in central-city population and workforce (U.S. Department of Transportation 1980b, 115). Retail activity trends in the urban core were not significantly different in beltway and non-beltway cities (U.S. Department of Transportation 1980b, 115). Suburban beltways have a small, negative, statistically significant effect on employment in wholesaling and service sectors of central cities; changes in industrial employment in central cities show a small but statistically significant negative relationship with presence of a suburban beltway after accounting for other factors (U.S. Department of Transportation 1980b, 115). The consultants concluded that:

The comparative statistical analysis shows that beltways can have small but significant effects on regional development patterns and the economies of central cities (U.S. Department of Transportation 1980b, 117).

The in-depth case studies of eight cities [Atlanta, Baltimore, Columbus, Louisville, Minneapolis/St. Paul, Omaha, Raleigh, and San Antonio] were chosen to include cities of different sizes and levels of economic health (U.S. Department of Transportation 1980a, 3-4). They showed varying effects. In Atlanta, a growing regional city, the beltway appeared to have attracted some retail and office space development, and garden apartments, to the corridor; it contributed to dispersal of economic activity but was not the major factor; it improved intraregional mobility, especially suburb-to-suburb, but was not a major factor in the region's growth (U.S. Department of Transportation 1980a, 7). In Baltimore, suburbanization had occurred but the beltway corridor's share had not increased since 1964 -- a clear indication that it had not affected overall demand; and the beltway's influence seemed to be waning, with far greater activity beyond it (U.S. Department of Transportation 1980a, 9). Columbus's beltway had attracted industry, but offices remained largely downtown (U.S. Department of Transportation 1980a, 10). In Minneapolis, "the tremendous suburbanization experienced in the Twin Cities over the past three decades cannot be attributed in any significant part to the presence of the beltway" (U.S. Department of Transportation 1980a, 12). In some cities -- Atlanta, Louisville, Raleigh, San Antonio -- the beltways appear to have influenced suburban development decisions, drawing high-density residential, commercial, and industrial activities into the corridors they serve, and reinforcing historic development trends; Atlanta recorded the most notable effects among large metropolitan areas (U.S. Department of Transportation 1980b, 97).

In most of the communities studied, planners recognized that highways could influence land use, but the policies they recommended in the 1950s and 1960s rarely sought to capitalize on the opportunities for corridor development, and hardly ever addressed the link between outlying development and downtown vitality (U.S. Department of Transportation 1980b, 94). Thus, rather remarkably, one of the greatest single opportunities in history to help shape urban land uses and activities was simply missed.

REFERENCES
American Association of State Highway Officials (1957) A Policy on Arterial Highways in Urban Areas. Washington, D.C.: A.A.S.H.O.

Automotive Safety Foundation (1964) Urban Freeway Development in Twenty Major Cities. Washington, D.C.: Automotive Safety Foundation.

Bolis, B. (1939) Le Autostrade Italiane al 1924-II al 1939-XVII. Le Strade, 21, 574-584 and 614-623.

Brodrick, A.H. (1938) The New German Motor-Roads. Geographical Magazine, 6, 193-210.

Buchanan, C.M. (1970) London Road Plans 1900-1970. (Greater London Research Intelligence Unit Research Report No. 11.) London: GLC.

Burnham, D.H. (1906) Report on a Plan for San Francisco. San Francisco: City of San Francisco. Reprinted 1971: Berkeley: Urban Books.

Cement and Concrete Association (1936) The Motorways of Germany. London: Cement and Concrete Association.

Cement and Concrete Association (1938) Roads Abroad. London: Cement and Concrete Association.

Charlesworth, G. (1984) A History of British
Motorways. London: Telford.

County Surveyors' Society (1938) A Scheme for
Motorways. London: County Surveyors' Society.

Drake, J., Yeadon, H.L., Evans, D.I. (1965)
Motorways. London: Faber and Faber.

G.B. Advisory Committee on Trunk Road Assessment
(1977) Report. (Chairman: Sir George Leitch.)
London: HMSO.

G.B. Road Research Laboratory (1948) German Motor
Roads 1946. (Road Research Technical Paper No. 8.
British Intelligence Objectives Sub-Committee
Overall Report No. 5.) London: HMSO.

German Roads Delegation (1938) Report upon the
Visit and its Conclusions. London: German Roads
Delegation.

Germany, Generalinspektor fur das deutsche
Strassenwesen (1938a) Fünf Jahre Arbeit an den
Strassen Adolf Hitlers. Berlin: Volk und Reich
Verlag.

Germany, Generalinspektor fur das deutsche
Strassenwesen (1938b) Die Reichsautobahnen:
Principles of Design. Construction and Traffic
Control. Berlin: Volk und Reich Verlag.

Guliano, G. (1986) Land Use Impacts of
Transportation Investments: Highway and Transit.
In: Hanson, S., op. cit., 247-279.

Hanson, S. (ed.) (1986) The Geography of Urban
Transportation. New York: The Guildford Press.

Hart, D.A. (1976) Strategic Planning in London:
The Rise and Fall of the Primary Road Network.
Oxford: Pergamon.

Hass-Klau, C. (1990) The Pedestrian and City
Traffic. London: Belhaven.

Jones, D.W., Jr. (1989) California's Freeway Era in
Historical Perspective. Berkeley: University of
California, Institute of Transportation Studies.

Jones, S. R. (1982) An Accessibility Analysis of
the Impact of the M25 Motorway. (TRRL Laboratory

Report 1055.) Crowthorne: Transport and Road Research Laboratory.

Lewis, H.M. (1949) Planning the Modern City. Volume One. New York: John Wiley.

New York: Regional Plan of New York and its Environs (1929) The Graphic Regional Plan: Atlas and Description. (Regional Plan, Volume One.) New York: Regional Plan of New York and its Environs.

Rayfield, F.A. (1956) The Planning of Ring Roads with special Reference to London. Proceedings of the Institution of Civil Engineers, Part 2, 5, 99-135.

Rose, M.H. (1979) Interstate: Express Highway Politics 1941-1956. Lawrence: The Regents Press of Kansas.

Schwartz, G.T. (1976) Urban Freeways and the Interstate System. Southern California Law Review, 49, 406-513.

Starkie, David (1982) The Motorway Age: Road and Traffic Problems in Post-War Britain. Oxford: Pergamon.

United States, Congress (1939) Toll Roads and Free Roads: Message from the President of the United States Transmitting a Letter from the Secretary of Agriculture, Concurred in by the Secretary of War, Enclosing a Report of the Bureau of Public Roads, United States Department of Agriculture, on the Feasibility of a System of Transcontinental Toll Roads and a Master Plan for Free Highway Development. (76th Congress, 1st Session, House Document No. 272.) Washington, D.C.: Government Printing Office.

United States, Congress (1944) Interregional Highways: Message from the President of the United States Transmitting a Report of the National Interregional Highway Committee, Outlining and Recommending a National System of Interregional Highways. (78th Congress, 2d Session, House Document No. 379.) Washington, D.C.: Government Printing Office.

United States, Department of Commerce, Bureau of Public Roads (1955) General Location of National System of Interstate Highways: Including all Additional Routes at Urban Areas Designated in

September 1955. Washington, D.C.: Government
Printing Office.

United States, Department of Transportation, Federal
Highway Administration (1976) America's Highways
1776-1976: A History of the Federal-Aid Program.
(Stock No. 050-001-00123-3.) Washington, D.C.:
Government Printing Office.

United States, Department of Transportation and
Department of Housing and Urban Development (1980a)
The Land Use and Urban Development Impacts of
Beltways: Case Studies. Prepared by Payne-Maxie
Consultants and Blayney-Dyett, Urban and Regional
Planners. (DOT-P-30-80-31.) Washington, D.C.:
Government Printing Office.

United States, Department of Transportation and
Department of Housing and Urban Development (1980b)
The Land Use and Urban Development Impacts of
Beltways: Final Report. Prepared by Payne-Maxie
Consultants and Blayney-Dyett, Urban and Regional
Planners. (DOT-39-80-38.) Washington, D.C.:
Government Printing Office.

Zapatka, C. (1987) The American Parkways: Origins
and Evolution of the Park-Road. Lotus
International, 56, 96-128.

Discussion

J.M.H. KELLY, Rendel Palmer & Tritton, London
To illustrate the positions of the Abercombie D and E rings and the M25 as mentioned by Professor Hall, I would like to show a plan (Fig. 1) pointing out the switch from the D to E ring in the Darenth Valley and the reverse near the A1(M).

As Professor Hall referred to the purpose of orbital motorways, I thought it worth showing the purpose of the M25, i.e.

(a) to provide a route around London by linking main radial routes
(b) to remove traffic particularly HGVs from unsuitable roads in and on the edge of London
(c) to link the South East's four main airports (Heathrow, Gatwick, Stansted and Luton)
(d) to improve routes from the Midlands and the North to the Channel ports.

I feel that it is fulfilling much of its purpose, although I doubt that much of the internal to internal radial movement on the edge of the urban area is occurring via the M25, particularly where the motorway is situated well into the green belt.

J.P. ROUX, Department of Transport, Pretoria, South Africa
What measures do you recommend to counter noise pollution, at the planning stage and on existing freeways around which residential areas have developed and concern is expressed regarding levels of noise?

S.N. MUSTOW, Consultant, Sutton Coldfield
With respect to the relationship of orbital planning and land use, if Abercombie had too static a view of the orbital as a cordon sanitaire, how could a more creative and dynamic relationship be achieved?

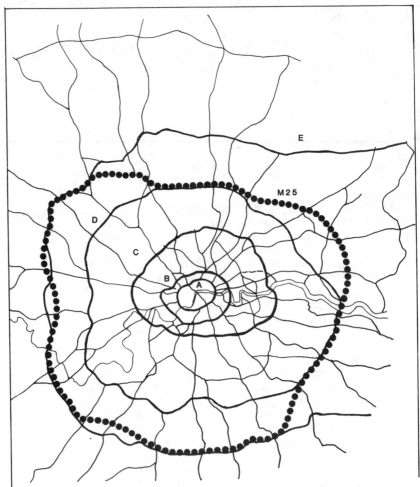

Fig. 1. As-built route of M25 compared with the five Ring Road plan set out by Abercombie in 1944.

R. CATHCART, Transportation Planning Associates, Birmingham
Is integrated planning, or perhaps a consensus on a strategic direction, simply unattainable in the UK given the national background?

M. SULLIVAN, Council for the Protection of Rural England
Peter Hall has explained in a fascinating way how the pre-war German autobahns were pursued. I hope that he will look at the near-empty, unaltered Hitlerian autobahns in Silesia, now in Poland, and one day visit what none of us have seen, the lost, derelict autobahn in former East Prussia now in the USSR (Kaliningrad enclave): an archaeological site to explore.
    The setting of cities varies greatly. What can be

an acceptable orbital around one city would do grave
damage to the setting of another of similar size.
In Britain, the green belts are maintained to
preserve an open background to a city, and one
cannot ignore the fact that this has been to
preserve countryside and villages - in which many
people who work to protect green belts live, and
those from the city seek to enjoy. Some of the M25
is not bad in local impact terms, but in cases such
as Swanley-Sevenoaks it should clearly have been put
in a tunnel part of the way. The Abercombie plan
for London shows this well.

Other cities in Britain present a mixture of
effects: Manchester has a mainly urban alignment, as
has Newcastle (Western Bypass), while Edinburgh's
periphery is less attractive than the city itself.
The West Midlands is the biggest conundrum: a
northern orbital does not pass through much
attractive countryside, but a western orbital would
be disastrous in planning terms: the Black Country
runs straight into rolling hill country (A.E.
Houseman's countryside) which stretches to Wales,
and dreadful developmental effects would ensue which
have not been considered. The southern orbital
(M42) shows such results in Solihull already. Why
not make better use of the M5 after taking M6
traffic off the urban section?

Overseas cities such as Atlanta (pinewood setting)
or Houston (flat plain) can accommodate orbitals
without such effects. Interestingly, some smaller
cities orbitals are of a lower standard than the
interstates which traverse the city (Austin, Texas;
Montgomery, Alabama), and so tend to be industrial
distributors.

At the other extreme, orbital routes around the
west of Geneva and south and west of Zurich are
being tunnelled to a considerable extent to meet
environmental objections. This is going to become
more and more necessary in Britain; its forthcoming
application in the Paris region is to be welcomed.

A.H. CRAIG, Babtie Shaw & Morton, Glasgow

Does history tell us that it is a good thing to plan
orbital motorways in advance of evidence of demand?
As we have heard, these motorways are tending to be
more and more expensive: one wonders what strength
of argument can be made to justify the expenditure
to the public.

PROFESSOR P. HALL

Mr Kelly's point about the relocation of the M25 is
well taken. As to the pattern of radial-to-radial
movements, I think that there is a good deal of

'dog-legging' from poor to good radials, as from the A3 to the M3 and M4, hence the notorious congestion on that sector. But only a detailed origin-destination survey could prove this.

In reply to Mr Sullivan, I have not been able to penetrate as far as Kaliningrad, but I have traversed the Polish section of the Wroclaw–Berlin Autobahn, which progressively degenerates from motorway to dual carriageway and then to single carriageway; I decided to abandon it when I encountered a Red Army soldier and his wife pushing a pram on the fast lane. This illustrates, I think, the enormous job that now needs to be done to restore the infrastructures of Eastern Europe up to Western standards, the scale of which the West German government has already begun to appreciate.

I take the point about the variation in setting as between one city and another. More should have been done, and more could still be done, to put orbital motorways underground in sensitive areas. The French did it 20 years ago with the Périphérique under the Bois de Boulogne, and now with the A86 near Versailles. With the exception of the M25 under Epping Forest, we have been almost unbelievably insensitive on this point.

In response to Mr Craig, yes, surely it pays to plan any motorway in advance of demand. James Drake is said to have planned the Lancashire motorways in the mid-1930s, when still Borough Surveyor of Blackpool. The exact lines were ready to put into the first Lancashire Development Plan of 1951. As a result motorways in Lancashire never proved really controversial, and Drake was able to produce a comprehensive network relatively quickly: as anyone visiting North East England today can see.

In response to Mr Cathcart: difficult, I think, but not impossible. The time will soon come when the consequences of failure to plan will be so grievous that – just as in the early 1960s – there will again be an espousal of comprehensive, integrated planning.

The operational control of motorways

J. WOOTTON, BSc (Eng), MEng, FCIT, FIHT, FBCS, Wootton Jeffreys Consultants Ltd

SYNOPSIS. Motorway control is aimed at optimising the use of existing motorways. The techniques used include speed control, access control, route information and warnings of hazards. The benefits are significant reductions in delay, fewer accidents and improved driver behaviour. Future technology will provide two-way communication with vehicles and improved means of controlling vehicles.

THE CHARACTER OF ORBITAL MOTORWAYS

1. Orbital motorways are enigmatic. They lead no where, yet they carry large volumes of traffic. Their main purpose is twofold. They permit circumferential traffic movements and provide a high quality bypass.

2. The ease with which circumferential movements can be made is important, for orbital motorways provide the only high capacity circumferential traffic routes in many major cities. This is the case in London, Madrid, Moscow, Paris and Washington, all of which grew rapidly along the radial routes provided by railways and where circumferential traffic movements have only been satisfied by the street patterns that have developed subsequently.

3. In some cities the roads are known by apt descriptive names. The "Almendra" in Madrid, the "Boulevard Peripherique" in Paris and the "Beltway" in Washington. Sometimes they are known more plainly. The M25 around London. But most have one factor in common. They are congested. Most of the time they provide a fast, attractive route to any alternative, but in peak hours, or when an incident occurs, long queues can form and delays are experienced. Even then the motorway can be a faster route to the alternatives.

4.   In the UK we often speak as though the congestion we experience on our motorways is unique. This is not so. I have known the Washington Beltway for over 20 years. It is a four lane dual carriageway yet for many years the peak hour traffic levels of congestion have been equal to those currently experienced on the M25. Similarly the Boulevard Peripherique in Paris have levels of congestion that match the M25. A flight connection involving a journey from Charles de Gaulle airport north of Paris to Orly in the south carries just as much risk of missing the flight as a journey from Heathrow to Gatwick.

FUTURE TRAFFIC DEMAND

5.   All the evidence suggests that the current levels of demand that cause the congestion on our roads will not subside. Vehicle ownership is set to continue to increase at over 2% per annum, while vehicle use is increasing at a 3% per annum. In the case of orbital motorways two other factors are important. There is no other form of transport that can provide the capacity that matches the demand for circumferential movement and, as with all other forms of transport, orbital motorways are in the process of changing the shape of our cities.

6.   The point concerning capacity may seem controversial, but circumferential movements tend not to have a collective focus and require a transport system that provides flexible routing and imply personal mobility. The pattern of streets and roads provides this flexibility, whereas collective transport systems, for example bus, train, airlines and freight delivery systems, are characterised by hub and spoke operations and are most successful where there is a focus to the demand.

7.   The development of any new transport facility provides people with the opportunity to live at more convenient locations and provide businesses with the opportunity to locate where they can obtain their resources and serve their markets more economically. The change in the patterns of location do not happen instantly but over a period of years. This is a timescale which humans find it difficult to comprehend and the changes are best observed as history rather than forecast.

8.   The full impact of the changes in land use pattern resulting from the construction of high quality orbital roads has yet to be seen, but there are already many examples of the impact they have. The location of high technology industries along Route 128 around Boston has existed for over 25 years. The office development and

retail centres at Tysons Corner, at the junction of the Washington Beltway and Dulles Airport Access Road, is now so large that it competes strongly with the centre of Washington. The same effects can be observed on our own M25. For example, Esso have moved their European Headquarters to within a mile of Junction 9 on the M25. At this same junction a new industrial site houses Dunlop, B & Q have constructed a superstore and over a hundred new houses have been built on land severed by the road.

9.   The consequence of these many factors is that traffic demand on orbital motorways will remain very high whatever measures are taken. They can be expected to be operating close to their capacity for many hours of the day. Any relief from new construction is likely to be temporary and the problem of controlling the traffic using any motorway is likely to remain of paramount importance.

THE OBJECTIVES OF OPERATIONAL CONTROL

10. The control of traffic using motorways has several aims. It is first aimed at optimising the use of the existing capacity of the road and as a rule of thumb will usually start to be beneficial and traffic flows exceed a 1200 vehicles per lane per hour. The better utilisation of the existing capacity of the motorway comes from detecting incidents, smoothing the flow and rerouting the traffic to other roads.

11. The operational control is also aimed at providing higher levels of safety. Smoothing the flow helps reduce the likelihood of accidents and, by providing warnings, the likelihood of secondary accidents is reduced.

12. The collection of traffic data and the continual monitoring of the state of the system also makes the task of the enforcement and operating agencies easier. Aids can be provided which will enable roadworks to be carried out more quickly and the rapid detection of incidents can lead to prompt action and a reduction in delays.

13. Success in achieving these objectives is obtained by the continuous collection of information. The rapid interpretation of the information results in actions to control the traffic directly or provide information to which drivers can respond. It is most important that any information that is provided is correct since it otherwise loses its credibility and hence its effectiveness.

SMOOTHING THE FLOW

14. There are few drivers who have not experienced a breakdown in traffic flow on a motorway. The incident that causes the breakdown in traffic flow can be many and varied, but it invariably occurs at high flows when vehicles are moving fast and

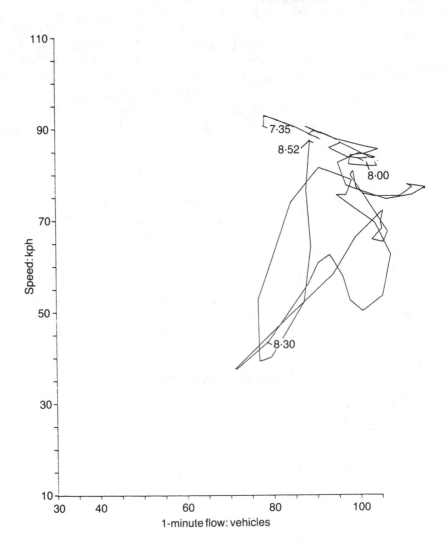

Fig. 1.   Minute-by-minute speed/flow observations south of the M6 Junction 10 before the introduction of access control.

an incident, perhaps trivial, causes one vehicle to brake. The result is progressively harder braking by the following vehicles, with the result that a shockwave travels back along the stream of traffic even causing traffic to halt. Once breakdown has occurred the resulting slow moving queues may last the whole duration of the peak period and drivers will experience poor driving conditions and considerable delay.

15. Figure 1 shows the effect of the breakdown of flow from speed flow measurements taken on the M6. Until shortly after 8 a.m. the flow of traffic was high at high speed. When breakdown occurred a queue formed and stop go conditions were experienced. The high speed and flow recovery does not occur until three quarters of an hour later at 8.52 a.m.

16. The breakdown of traffic flow can be avoided or at least delayed if the shockwaves resulting from an incident can be prevented from propagating through the traffic stream. The purpose of motorway control and surveillance systems is to achieve this end. The techniques used include speed control, ramp control and diversion routing.

MOTORWAY CONTROL AND SURVEILLANCE

17. One of the earliest systems to employ the techniques of motorway control is the centrally integrated traffic control system (CITRAC) introduced by Strathclyde Regional Council. The system includes surveillance of traffic on the motorways, ramp control and diversion routing as well as being integrated with the urban traffic control system of Greater Glasgow (ref. 1).

18. Actions are taken by a duty police officer in the central control room. The police officer enters "primary restrictions" which include speed advisory information as well as information on emergencies, operational restrictions for roadworks and speed restrictions associated with bad weather. The central computer automatically generates secondary restrictions which are designed to ensure safe driving. The monitoring of the system is through closed circuit television, patrol cars and messages phoned in from the roadside. The system is designed to be inherently safe so that speed changes are phased over time and no two adjacent gantries should show a speed difference of more than 20 miles per hour. It is claimed that as well as smoothing the flow, the system has resulted in a 20% reduction in off peak traffic accidents, particularly those concerned with excessive speed and lane changing.

19. CITRAC also controls the traffic leaving the motorway and signs a diversionary route when an alternative to the motorway is less congested. The traffic leaving the motorway is controlled by signals at the end of the exit ramp and diversionary routes are given by special signs before the motorway is joined. The system does not use ramp metering to control traffic entering the motorway, though this is intended in a future stage of development.

SPEED CONTROL

20. Some of the most interesting results of speed control experiments have been reported by Jenezon et al (ref. 2, 3), who have worked for 15 years in Holland on motorway control and signalling systems. The systems discussed comprise gantries over each carriageway with matrix signals over each lane. A matrix signal can show a speed, a red cross, a change lane arrow or an end of temporary limit sign. The information on the gantries is controlled by out-stations which in turn are connected to a central control facility. Traffic flow and speed measurements are made by loop detectors set into each lane of the carriageway and this information is fed to the central control system.

21. Although the technology of the Dutch system is interesting, the most important lessons are learnt from their experience. For example it is reported that the benefits to be obtained from dynamic traffic monitoring and central control are more than four times those of non-dynamic systems.

22. The spacing of the gantries and detectors are also important. There is no point in warning traffic against hazards which are more than 1.5 kilometres away because the warning will be forgotten after a minute. Gantries are best placed no more than 500 metres apart.

23. The rapid and accurate detection of traffic behaviour and incidents is also important. Detectors spaced 150 metres apart will not miss an incident, but there is no advantage in spacing detectors more closely. The preferred spacing for detectors is about 500 metres and Jenezon et al report that with longer distances the relationship between the effect an incident can have on traffic and the behaviour measured at the detector rapidly becomes vague.

AUTOMATIC OR MANUAL CONTROL

24. Perhaps the most telling comment from the Dutch work is that the computer works rather better at operational control than a human operator. When an operator was required to verify the

computer recommendation, a time delay of up two minutes occurred between the moment a new traffic measure was proposed by the computer and the moment the signs were actuated on the gantries. Delays of this length produce a negative response from road users who, after 2 minutes delay, will be experiencing the consequences of the incident. Furthermore, there was no single case where the operator could have done better than the computer. The system is now fully automatic without any operator intervention. The result is that the automatic system takes action to prevent queues on average seven times more than a human operator. It is concluded that these prompt actions have contributed significantly to a reduction in accidents.

25. Experience suggests that drivers do not obey speed directives particularly well. This is particularly critical in fog conditions. The evidence presented by the Dutch suggests that driver behaviour improves and becomes much more responsible where the information is up to date and immediately seen to be relevant. Using the automatic incident detection mechanisms to warn drivers of queues or slow traffic installs discipline into the drivers behaviour. This technique is used in foggy conditions and is believed to have contributed to a significant reduction in accidents in fog.

ACCESS CONTROL

26. The first experiment in access control in the United Kingdom was implemented on the M6 on the southbound entry slip road of Junction 10 (ref. 4, 5). The success of the system can be seen by comparing figures 1 and 2 with free flow conditions persisting when the access control was used (fig. 2). The success has resulted in access control systems being installed also at Junctions 9, 7 and 5.

27. Prior to the installation of the access control system at Junction 10 congestion was experienced for regular and prolonged periods of time. Congestion was caused by breakdown in flow in the section of road downstream from Junction 10 and would often last for two hours or more with tailbacks extending upto 10 kilometres. The delays arising from the congestion were estimated to be in the region of 100,000 vehicle hours per year.

28. The flow breakdown was found to be related to short term peaks in flow through Junction 10. With a significant volume of traffic gaining access to the M6 at Junction 10 the junction made an ideal test site. The access control solution uses conventional traffic signals, with standard minimum stage lengths and amber periods, to control the traffic joining the motorway from the Junction 10 slip

road. The signal phases are controlled by algorithms which respond to the speed flow information obtained from loop detectors located on the slip road and in the main carriageway. The signals remain on green unless there is need to control the slip road traffic. This need occurs when the combined slip road and main carriageway demand flows exceed a critical flow limit or when the traffic speeds downstream of the merge fall below a predetermined threshold. When either of these conditions occur a red signal is shown to the slip road traffic.

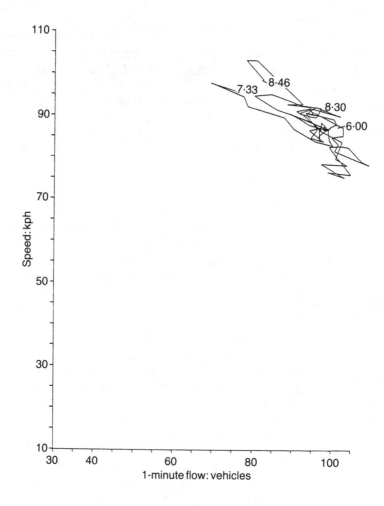

Fig. 2. Minute-by minute speed flow observations after the introduction of access control at M6 Junction 10

29. The flow limits and speed threshold vary with the prevailing weather conditions. The M6 runs south east-north west between Junctions 9 and 10 so that the low sun early on a winter morning induces slower traffic speeds as do wet driving conditions. Consequently a database must first be built up to determine the thresholds and the operational control must know which thresholds apply on a particular morning. This is achieved automatically by loop detectors which continually monitor the speed and flow of the traffic.

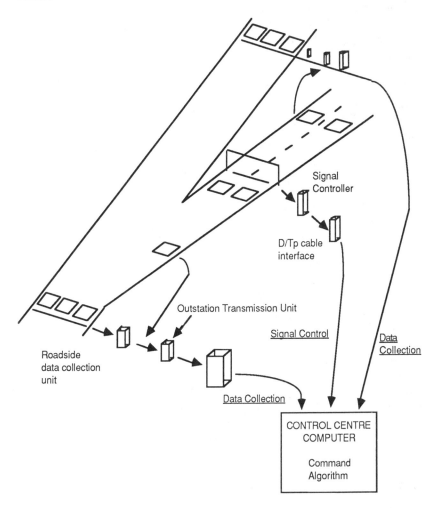

Fig. 3.   Diagrammatic layout of the access control system at Junction 10 on the M6

30. Figure 3 shows diagrammatically the layout of the access control system at Junction 10. The access control signals are mounted on a gantry over each lane of the slip road about 250 metres from the entry to the slip road. Drivers are warned by variable message signs located at the entry to the slip road that the traffic signals are in operation.

31. The success of this system, as with other motorway control systems, is in the reduction of the delays that occur. The flows through the bottleneck section downstream of Junction 10 have increased by 3-5%, but the reduction in traffic times have averaged more than 20%. A cost benefit analysis has suggested that the first year rate of return was at least 20% and perhaps as high as 40%.

32. A matter of initial concern was whether an access control system would have any impact on the adjacent all-purpose road networks. The evidence that has been collected suggests that a small proportion of drivers, no more than 5%, were deterred from using the slip road at Junction 10 but that any change has not had any significant influence on the traffic movements on the adjacent all-purpose road network.

COMPONENTS OF MOTORWAY CONTROL SYSTEMS
33. The basic components of an integrated motorway traffic control system are:

> A real time database
> Interpretation of the data
> Information feedback and control measures

34. The real time database should be constructed from real time speed and flow measurements at key locations. These will usually be obtained from loop detectors. The detection of incidents is also important and the speed flow measurements will help in this. Other methods include image processing and an example of this is the IMPACTS system developed by the Transport Studies Group at University College London. IMPACTS can observe a section of road and decide whether traffic is moving or stationary. This permits the rapid detection of incidents and allows the computer system or an operator to act quickly on an adverse event. Other information which should be included in the database will include weather conditions, special events, emergency facilities, roadworks, etc.

35. The interpretation of the data is crucial. The control system must react rapidly to the change of conditions that occur and make decisions either through an operator or automatically. At present control is usually exercised through a control centre where staff are

trained to understand traffic behaviour and the policing aspects of traffic control. In future, the systems are expected to become more automatic and the Dutch experience suggests this will have added benefits.

36. The control system will fail unless the information that is passed back to the users is reliable, accurate and relevant. Drivers needs are likely to be different and information must be structured to ensure so that drivers are not forced to take advice that they view as unreasonable.

METHODS OF PROVIDING INFORMATION

37. The methods of supplying information to drivers are various and new methods are likely to develop in the future. At present variable direction and message signing is of most importance. The information that can be of value to drivers includes warning them of restrictions in capacity, informing them of potential hazards, and advice on the route they should follow.

38. The police and traffic authorities distribute traffic information to radio stations, but this is usually broadcast at intervals which are not suitable for real time control. Because of the delays, information transmitted by radio is most likely to be of benefit prior to the trip commencing or where the incident is some miles ahead on the journey being made.

39. The greatest benefit is likely to be obtained from a traffic control system when the information can be given personally to the driver within the vehicle. Such systems are now in the process of development and installation. For greatest success they depend upon a two-way communication system so that the driver can be informed and his response is monitored. The AUTOGUIDE system which is in the process of being installed in London will provide many of the facilities required by the third component of a real time traffic control system.

FUTURE TECHNOLOGY

40. The DRIVE and PROMOTHEUS research initiatives are both leading to technical developments which have a major impact upon the way we use vehicles and will control traffic in the future. In the period from 1995 to the year 2000 the use of at least five new systems are likely to be common throughout Europe.

a) A new standard digital communication system (GSM) will be installed throughout Europe. This will provide high capacity two-way communication by mobile telephone. The system will have the capacity to provide every vehicle with its own communication channel to a control centre.

b) Broadcast traffic messages via FM radio will be more local and hopefully advice will be given more rapidly than at present. The radio data system (RDS) is already part of the new standards for FM radio and within this system the traffic message channel (TMC) will allow the radio to be interrupted so that relevant messages can be given to the driver.

c) Vehicles will become available with intelligent cruise control. The vehicles will be able to brake as well as accelerate to a predetermined speed and recognise vehicles or objects that are close to them. The vehicle manufacturers already have test vehicles that can follow one another at very short distances apart, but one has to wonder what will be the reaction of drivers travelling at 60 miles per hour 10cm from the vehicle in front.

d) Interactive route guidance, of which AUTOGUIDE is the most immediate example, will be installed in most of our major cities. These systems, whether they use existing infra-red transmission systems or the digitial cellular radio provide an important interface for communicating with the driver.

e) Automatic toll collection systems are already in existence and it can be expected that demand will be to some extent controlled by the payment of fees for the use of roads.

41. All of these new systems, and others that evolve from them, will allow the more efficient use and control of the existing road space. It is difficult to calculate the increase capacity that will result but it seems likely that an increase in capacity of the existing road space of $1/2$% to 1% per annum is possible for the foreseeable future.

42. Technology will also have an impact on safety. There will be aids to assist and warn the driver of incidents and hazards. Systems may also be in use that will determine whether the driver is fit to drive.

2020 VISION OR "THE 2 SECOND SLOT"

43. Given the new developments in technology discussed in the preceeding paragraphs, the continued growth in traffic demand and the unlikelihood of being able to satisfy the demand by new road construction, it is not unreasonable to speculate on the way in which a journey might be made by personal transport in the year 2020. The journey might progress as follows:

i)   The driver enters the car and dials the location of the destination. He waits.

ii)   Five seconds later he's told the journey can begin in two minutes time.

iii) Two minutes later the central computer allows the car ignition to be turned on. The driver commences the journey.

iv)   Throughout the journey the intelligence built into the vehicle controls its speed and position relative to other vehicles and the central computer monitors traffic conditions and redirects the driver so as to avoid incidents.

v)   The driver arrives safely at the destination.

vi)   The bill is received for the journey one month later.

44. Whether the traffic control systems exist in this form is sheer speculation, but it is important to note that the digital cellular radio provides the necessary two-way communication system, AUTOGUIDE already provides the necessary user interfaces and the PROMETHEUS and DRIVE initiatives should produce the traffic and vehicle control systems that are necessary.

INVESTMENT IN TRANSPORT

45. Capital investment is the key to the growth in productivity and income for a nation or indeed a city. This means that the investment in additional transport facilities should be aimed at providing increased productivity through more efficient transport systems. Historic examples of the value of such investment include the canals, railways, and more recently air transport. The question is where should capital investment in transport be directed in the future in order to maintain and improve the quality of life, that most people desire through economic growth, a better environment and safer travel.

46. There can be little doubt that we cannot expect to continue to provide additional roadspace to satisfy the uncontrolled demand of existing vehicles and the increase expected in the future. Historically it has been technology that has led the development of new transport systems and it seems that this should be the priority for transport investment in the future.

47. The technology that is developing rapidly is in the field of electronics and communications and these have yet to be applied fully to our transport systems. It is the authors view that a priority for future investment is on the application of these technologies to transport systems, with the result that new forms of traffic control will come about and new forms of collective transport systems, offering greater personal flexibility, maybe developed.

REFERENCES

1. MOWATT A.M. and YOUNG A.D. CITRAC - the first five years. Traffic Engineering and Control, Vol. 25, No. 5, May 1984, 251-259.

2. JENEZON J.H., KLIJNHOUT J.J. and LANGELAAR H.C.G. Motorway Control and Signalling. Traffic Engineering and Control, Vol. 28, No. 6, June 1987, 349-355.

3. REMEIJN H. The Dutch Motorway Control and Signalling System. Rijkswaterstaat, Traffic Engineering Division, TXR-S 131/103/150.

4. RIDOUT G. Control of Access Eases Motorway Congestion. Surveyor, 26th June 1986.

5. OWENS D. and SCHOFIELD M.J. Access Control on the M6 Motorway : evaluation of Britian's first ramp metering scheme. Traffic Engineering and Control, Vol. 29, No. 11, December 1988.

6. KEEN K. Traffic Control at a Strategic Level. Second International Conference on Road Traffic Monitoring, The Institution of Electrical Engineers, 7th-9th February 1989.

The M25 London orbital motorway – a case study

M. SIMMONS, MA, DTP, MRTPI, London Planning Advisory
Committee

INTRODUCTION

1.   The M25, 119 miles around and enclosing the London
conurbation, a population of some 7.4 million and containing
3.8 million jobs, was completed in 1986 after 13 years
piecemeal construction. It is the only full orbital route in
the London region and its western half is already carrying
traffic, 30 to 70% in excess of planned capacity. This paper
considers its role and significance in the region's
transportation and development structure and explores the
proposition that the motorway has been narrowly conceived,
without adequate regard to either its role in the total
requirement for orbital movement in and around London, or its
impact on and relationship with land use strategy for the
London region.

2.   The second section of the paper examines the origins of
the motorway, comparing its place in plans for Greater London
with its coming together and implementation by the Department
of Transport during the 1970s, constructed to a standard and
with a pattern of intersections which can at best be called
pragmatic.

3.   The third section considers the traffic and transport
situation now facing the M25, comparing recently recorded
traffic flows with pre-construction forecasts. It notes the
criticism of Department of Transport procedures, particularly
its failure to assess generated traffic, and the concern
expressed over the situation, particularly on the western side
of the motorway. The Department's response is commented on,
including the options put forward by Consultants and the
proposals made in the latest Roads White Paper. This section
concludes with the different needs for orbital movement, as
the basis for considering alternative provision.

4.   The fourth section relates the M25 to the London
region's development, in a situation where development has
always followed transport infrastructure. It traces the
relationship through the slow expansion of regional planning
activity over the last decade, beginning with SERPLAN's
examination of the development impact of the M25, through
references in Regional Planning Guidance for the South East,
up to SERPLAN's current review of regional strategy for the
Secretary of State for the Environment.

Orbital motorways. Thomas Telford Ltd, London, 1990.

5.   The fifth section continues to examine the alternative
options for orbital movement, related to the changing
structure of land use activity in london and the South East.
It notes how the Department of Transport's proposals for an
east-west route across the north of the region, together with
a Lower Thames Crossing, and furtherance of schemes in the
Eastern Thames corridor, would relate to where capacity for
development exists in the region. In considering the problem
sector of West London, however, the question is whether demand
for road movement can and should be met, or whether the future
should lie in public transport improvements, noting that this
is the outcome of the London Assessment Studies and the
potential shown by a light rail scheme centred on Croydon.
6.   Finally, a concluding section points to the elements of
an integrated strategy which would link the functions and
requirements of orbital movement to the development strategy
for London and the South East. It suggests that, beyond
widening to 4-lane, there should be no further on- or near-
line improvement to the M25.

GENESIS OF THE M25, IN THE LONDON REGIONAL CONTEXT
7.   The concept of an orbital motorway around London first
appeared formally in Abercrombie's seminal Greater London
Plan, 1944 (ref. 1). This put forward a simple, coherent
framework of "express arterial roads" made up of ten national
or regional radial routes and two rings. The two strategic
ringways (of five in all) comprised one close in, around
London's central commercial and industrial core; this would
constitute the inner end of the radials, and would avoid
traffic passing through and creating congestion in Central
London. The outer ring was designed to circumscribe the
built-up area of the conurbation, some 20 km (12.5 miles) on
an average from the centre. This "D ring" would have been 147
km (92 miles) long; beyond it extended the Green Belt,
although the boundary between built up London and Green Belt
was, and remains irregular. Abercrombie described its purpose
thus: "there is a need for another ring, close to the
periphery of the built-up area, for express and through
traffic;: this is a major topographical deficiency in London's
road system". It would provide a "magnified bypass" providing
the shortest route for regional traffic distribution between
the radials, without penetrating the suburbs. It would link
the suburban centres, Docklands, and airports. Beyond the D
ring, north and south orbital roads, sub-arterial in standard,
would provide a "parkway" route between centres in the Green
Belt ring; Abercrombie saw this as too far out to fulfil the
strategic, structural role. These routes were enshrined in
the first Development Plans prepared by the London County
Council and the surrounding Counties pursuant to the Town and
Country Planning Act 1947; they were identified and
safeguarded, but with little apparent impact on land use
provision, and no construction took place.
8.   A new plan for Greater London was prepared in the late

1960s following the creation of the Greater London Council in
1965. The Greater London Development Plan (GLDP), submitted
for Government approval in 1969 (ref. 2), retained
Abercrombie's basic radial and ring structure, but in a
modified form. Orbital movement was to take the form of three
Ringways: a "motorway box" around London's core; Ringway 2
(upgrading the existing North Circular Road plus a new route
through suburban South London), and Ringway 3. The latter was
Abercrombie's D ring, circling London, seen as "providing for
traffic which would otherwise penetrate deeply into the urban
area". Abercrombie's north and south orbitals were retained,
largely beyond the Greater London boundary in the Green Belt
to the south, west and north (not east) of London; their
construction was seen as "less urgent". The Plan's Report of
Survey stated: "the most urgent and critical necessity is the
linking of the national radial roads by orbital routes or
Ringways to serve the present and increasing number of
journeys not making for or leaving the centre" (ref. 3).
Priority was to have been given to the northern and western
sectors, from the A10 in the north via the M1, M40, M4 and M3
to the M23 in the south, "where traffic generation is most
intense and needs to be related efficiently to the main
radials". Ringways 2 and 3 were seen as complementary: R2
distributing radial traffic around the growing "middle ring"
of London's employment areas, inwards from which the transport
emphasis would be on mass transit and restraint of car use.

9.   At the end of the 1960s the Government established a
Joint Planning Team with local government to prepare an
overall regional plan for the South East. This reported in
1970 (slightly later than the GLDP) and consciously sought to
relate motorway planning to a regional development strategy
looking 20 years ahead. This assumed that Ringway 3 would be
constructed, but its basic perspective was to identify the
strategic road framework necessary for the South East to
function and develop efficiently, including that required to
stimulate the desired development strategy outside London.
This strategy sought to concentrate on future development into
designated "growth areas" beyond an extended Green Belt. This
led to proposals for two new part-orbital routes: one from the
M1 (in the Milton Keynes area, one of the new growth areas)
via the A1 at Stevenage to South Essex (another growth area),
and via a new Lower Thames Crossing to Kent (with two smaller
growth areas) and the Channel ports; the other from the M1
west of London to link with two major growth areas centred on
Reading/Basingstoke and Portsmouth/Southampton. A further
orbital link from these areas (served by the M3 and M4
radials) to Kent took the line of the old South Orbital, which
by now was being referred to as the M25, so numbered from the
old A road (from Guildford to Maidstone) it would replace.
The Strategic Plan for the South East (ref. 4) therefore
proposed outer orbital routes with quite distinct functions
from Ringway 3, related to growing demands for movement
between outer growth areas providing for the relocation of

economic activity, a component part of the regional
restructuring the Strategic Plan would provide for. In
recognising the link between strategic transport
infrastructure and the location of development, it reveals a
clear difference in the conception of new routes, from the
traditional Department of Transport process which is based
upon providing for the growth of existing traffic patterns,
through reassignment models.

10. The next stage in the saga was the Government's
decision on the GLDP, which did not occur until 1976, seven
years after the GLC submitted it. A Panel of Inquiry, under
FJB (later Sir Frank) Layfield QC was set up to examine the
Plan, and reported in 1973 (ref. 5). The proposed orbital
motorways was one of the key issues examined into, having
generated major opposition. The Panel accepted the need for
provision for orbital movement, both within London and around
the outer edge of the conurbation. Inside, the debate was
between a tight "motorway box", or the "middle" Ringway 2:
which would best canalise heavy traffic and improve the
environment of heavily built-up areas? A modified version of
the inner ringway was preferred; Ringway 2 was seen as too far
out from the main traffic-generating areas. However,
improvements to the North Circular and its extension across
the Thames in the east to the M20 were agreed. Outside, the
Panel accepted the need for a route to channel traffic which
would otherwise penetrate the built-up area, and to assist
movement between the outer suburbs, which were anticipating
traffic-generating growth. The Panel took a pragmatic attitude
to its route, given the volume of objection to Ringway 3,
particularly its southern and western sections. It noted the
degree of commitment by now apparent to the South Orbital
(M25) further out, and to parts of Ringway 3 in the north and
east (between the A1 and A10, and between the M13 and the
M20/M20 including the Dartford Tunnel). It therefore proposed
the omission of the south and west parts of Ringway 3 in
favour of the M25. It noted the need for further orbital
capacity to be studied in outer London, fearing (with
prescience) reliance on one orbital route.

11. In approving the GLDP in 1976, the Secretary of State
did not agree with the Inquiry Panel on the inner orbital
route, concluding that the cost and impact would be too high
(ref. 6). He partly reinstated Ringway 2 as the improved
North Circular, to be extended in East London across the
Thames (the "East London River Crossing") as far as the A2
Channel Ports radial, thus linking all the main national
radials from the A4 across North London. He recorded as "for
further consideration" the need for improved facilities for
orbital movement in South and West London, saying this could
include sections of both Ringways 2 and 3, and referred also
to studies of how to provide orbital movement around Central
London. After seven years' deliberation, therefore, difficult
decisions were shuffled off, to emerge again in the following
decade. The South East Economic Planning Council of the time

welcomed the Government's decision to support the M25 as
London's outer orbital on the grounds of ease of construction,
but sought improvements to orbital public transport to ease
labour problems by making it possible for workers to take jobs
further away from their homes (ref. 7).

12. During the 1970s, the preparation of what was to become
London's only Orbital Motorway had been proceeding through
Department of Transport's annual programme statements or Roads
White Papers. This process is only loosely associated with
Development Planning, until individual schemes come before
public inquiry, a disjunction which has led to much argument.
The Government had, however, become committed to the phased
construction of a nearly complete orbital by 1975. In
concept, this comprised two distinct routes. The first was
the "original" M25 - the old South Orbital - extending from
the M20 west of Maidstone passing Sevenoaks, Reigate and
Leatherhead to intersections with the M3 near Staines and the
M4 west of Heathrow Airport to the A412 north-west of Watford.
North and east of London the second proposed motorway was
conceived from the M1 (via the M10) to the A1 and then on the
line of Ringway 3 (but passing outside Romford rather than
R3's environmentally-sensitive route through Hainault Forest
and between Romford and Dagenham) to the fixed point of the
Dartford Tunnel and the established R3 line between the A13
and the M20/M20 at Swanley; this motorway was originally
referred to as the M16. The linking of the two routes, via an
outer route north of Watford and a section strongly contested
on need between Swanley and Sevenoaks in Kent, were not
determined until the early 1980s.

13. Construction commenced in 1973 and the first short
sections were opened in 1975/6. Further sections cleared the
planning process and entered the construction programme over
the succeeding decade. The complete orbital ring was open in
1986. The Government's 1983 Roads White Paper (ref. 8) gave a
typical progress report: during the financial year 1982/83, 18
miles had been completed (in the north-east between the M11
and the A13) so that 62 of the total 119 miles was open by
mid 1983, in seven separate sections. During 1983/84, the
section between Reigate and the M3 in the south-west would
open, and the remaining 6 sections were under construction or
in the preparation pool. The White Paper was by now able to
state: "the M25 is part of the national motorway network, and
is also an important link in the regional main road network as
a distributor for London traffic, and as a link between the
major airports of the London region. It is already providing
relief from congestion and environmental disturbance in parts
of outer London .... as it nears completion in 1986 these
benefits, as well as those for traffic and industry, will
become increasingly obvious". Far less obvious to the
Department of Transport was whether one orbital route could
possibly cater for this confusion of functions.

14. This review of the history of the London Orbital
Motorway saga has shown how a "grand design" degenerated into

the piecemeal implementation of one route, the M25, expected
to cater for all orbital movement around London. The
incremental nature of the decision-making process is
demonstrated by examining the number of intersections which
were constructed over the long period of the motorway's
construction. The pattern of intersections reflects, firstly,
a disjointed planning process negotiated between each county
area and the GLC and the Department of Transport, where local
interests and needs dominated any overall strategy and a
confusion of function and purpose set in; and secondly, the
Department of Transport's own incremental process of
programming and constructing the route over a 13 year period.
Including those on the Dartford Tunnel approach roads, the M25
contains 32 separate intersections; 9 are with the national
radial motorway network; 9 with other Government Trunk Roads
and the remaining 14 are with county roads serving more local
functions. The frequency of intersections are on average
every 3.75 miles (6 km). This allows much relatively short-
distance "local" use of the motorway for work or pleasure
trips, in addition to the longer-distance "London Bypass" use
as part of the national road transport system, which has led
to the much heavier than forecast traffic volumes on the
motorway now being experienced: "the M25 problem".

THE TRANSPORT AND TRAFFIC SITUATION: FORECASTS, FLOWS AND
FUNCTIONS
15. Traffic overloads and growing congestion on the M25 has
become a major issue since 1986 when the final sections closed
the gaps between the "South Orbital" and "Ringway3" elements
to create the complete London Orbital Motorway. This issue
found expression in the National Audit Office's examination of
the Department of Transport's new roads appraisal process in
1988, in which the NAO paid attention to the relative accuracy
of the Department's traffic forecasts, in particular to the
wide differences between the forecast and the actual flows on
the M25. Observed traffic flows in 1988 were above the design
year range for a dual 3 lane motorway on 21 of the 26 sections
of the M25; Appendix B shows that in the worst sections,
between junctions 12 and 16 in the west, flows of 123,000-
136,000 vehicles per day were recorded, compared with a design
flow of 79,000 anticipated 15 years after the motorway opened.
The whole western side of the motorway, between the A3 and the
M1, is carrying flows in excess of 100,000 v.p.d., whereas
flows in eastern sections, between the A12 (junction 28) and
A21 (junction 5) are in the range 60-80,000 v.p.d. Overall,
flows in the design year were expected to be 74% above the
original forecasts. It is apparent that as each new section
of the motorway opened, linking further radial routes, more
traffic was attracted. Quantum increases occurred as the
final links were completed; in the north-western quadrant,
traffic in the ten sections adjacent to the final link
increased by between 30 and 70% within 6 months of its opening
to traffic.

16. The NAO noted that the M25 forecasts were undertaken in the 1970s; they were built up from a base in the late 1960s (for the "south orbital") and revised, at a time when forecasts were over-predicting. The NAO considered how the users of a new route are abstracted in the DTp's forecasting model; this emphasises existing traffic, switching to the new route, the growth of which depends on assumptions about the macro-economy; this is known as "reassigned traffic". To this base-load, allowance is added for traffic which will change destination due to the new route, e.g. by travelling further ("redistributed traffic"); traffic switching from public transport; and people who did not previously travel attracted by the new possibilities ("generated traffic"); allowance is also made for known land use developments.

17. The NAO expressed great concern about the wide variations between the 1970s forecasts (on which the dual 3-lane standard and the pattern of intersections had been adopted) and the 1988 traffic levels, which "raise questions about the economy, efficiency and effectiveness with which resources were used in the construction...". Discussion centred on the extent to which the M25 was generating new, unforeseen traffic. The NAO considered that there was no proper assessment of both generated and redistributed traffic, and that "unexpected developments" were occurring. These clearly, in its view, were significant contributors to the unexpectedly high traffic flows being experienced in the western sectors. The NAO recommended that "induced land use change" should be given more emphasis in forecasting, although it did not press the conclusion that the impact of the motorway on the location of economic activity adds a whole new dimension to "generated traffic". Its report accepted that the M25 would need to be widened at very great and inefficient cost, and raised doubts as to whether a dual 4-lane standard would cope with vehicle flows likely to exceed 150,000 v.p.d.

18. The House of Commons Public Accounts Committee took up the NAO report in its 1988/89 hearings, and cross-examined the Department of Transport on its failure of the forecasts (ref. 10). The DTp defended its approach, claiming that the under-estimates could be explained by much faster growth in the national economy during the 1980s, giving rise to more traffic; it took the view that "the inclusion of a closer assessment of generated traffic would not have made much difference to the forecasts". This did not convince the Committee, who took the view that even in the buoyant south-east, general economic growth cannot fully explain congestion on the M25: "we recommend that DTp should reconsider the need to include assessments of generated and redistributed traffic in their future forecasts, and should take account of regional growth factors". "More account should be taken of the "wider and more strategic consequences." One member of the Committee, Michael Shersby MP, put the point that the M25 had achieved a great increase in accessibility, so that people were now able to visit new locations for both business and

pleasure; this change to travel patterns had led to an enormous increase in traffic. The Committee went on to express great concern about the costs and serious economic consequences arising from the worsening congestion, and suggested that certain access points should be closed to reduce short movements, particularly in the most congested western sections.

19.   The DTp's response is instructive: "The M25 functions as an urban motorway ... road movement in and about London is particularly difficult ... the problem of designing roads for a capital city where congestion is endemic". This indicates the inadequacy of the traditional forecasting process in a situation where the M25 creates a quantum increase in accessibility in a congested city-regional situation, releasing new demands for movement. There is no awareness, however, of the response to the new accessibility in the business, consumer and labour markets, and the traffic generation likely to result.

20.   The Government's mounting concern about traffic overloads led to the commissioning of a study by consultants Rendel, Palmer and Tritton which reported in the summer of 1989 (ref. 11) . Their report concluded that the high growth of demand for usage of the M25 would continue, unless restrained by capacity limitations which will suppress movement. Congestion will worsen and extend to longer periods of the working day, and will spread to the adjoining network. Traffic Management measures were recommended, aimed at "reducing and managing congestion, primarily by controlling peak access to the motorway". Close of certain intersections did not however find favour, due to the lack of alternative routes, although the matter should be kept under review. The Consultant's main proposal was to widen the motorway to dual 4-lanes; this was seen as "inescapable" and should begin now, with a trial of reduced lane widths as an interim measure.

21.   The Consultants also proposed that, if traffic growth is not to be suppressed, provision of further capacity should be considered in the wider context of the surrounding road network: an admission of the failure of relying on one orbital route. The options include improving parallel routes near the line of the motorway, as "collector/distributor" roads, and developing other routes both inside the M25 ring, and further out. Inside the ring, "the strategy of enhancing the road network near the fringe of the London conurbation has merit, especially in the heavily trafficked western sector ... to assist the M25 in carrying orbital movements ... for traffic having origins and destinations within the M25". However, the Consultants found little scope for this south of the Thames; while in the critical western sector, although a diversionary route presents itself (A312/4090/410, linking Feltham, Northolt, Harrow and Edgware), major investment would be necessary if it is to offer an effective alternative. The re-run of the GLDP debate is starkly evident!

22.   Beyond the M25 ring, while the Consultants found that

opportunities to enhance existing routes were not obvious, a study should be undertaken of the "outer orbital corridors". They referred to the potential for a northern orbital corridor to include a Lower Thames Crossing which would provide for through traffic to Stansted and Gatwick Airports, and the Channel Tunnel. A new orbital route could then link the M40 in Oxfordshire, the M1, the A1 near Stevenage, Stansted, the A12 near Chelmsford; a route 15-25 km outside the M25. The need for an outer western link between the M13, M4 and M1 was also referred to. The Consultants did not seem to realise that these were the selfsame corridors considered necessary in the 1970 Strategic Plan for the South East!

23. The Government's response is set out in its 1989 Roads White Paper, "Roads for Prosperity" (ref. 12). This accepts the proposal to widen the M25 to dual 4-lane standard, at an estimated cost of nearly £1 billion. Priority will be given to the western sections. Two further elements of this latest statement of Government road policy are of interest:-

- Phased improvements along an east-west corridor north of London, linking the M40 in Oxfordshire past Stevenage to the existing A120 leading to the A12 at Colchester, Ipswich and the ports of Felixstowe and Harwich. This will create a tangential route some 20 km north of the M25 linking the national radial routes to the West Midlands and the north.
- Commissioning of studies to examine the need for and the location of three further possible outer non-radial routes: a lower Thames Crossing (presumably to link the A12 with the the M2 and M20 routes to the Channel); a link between the M3 and M40 west of London (to relieve the most congested part of the M25), and in an east-west corridor south of London (to provide a route between Hampshire and Kent).

The implications of these are considered in the fifth section below (paragraphs 35-45).

24. This section of the paper is concluded by a summary of the different needs for orbital movement around London. The National Audit Office saw the M25 as:

- an inter-urban route, forming a fundamental part of the national motorway network for non-London traffic (the "London Bypass" function);
- a city-regional motorway channelling London-bound traffic around the outer edge of London, providing the most appropriate access points to the London conurbation and near it (including the Airports);
- the primary urban distributor for the Greater London area, with all the attributes of an urban route for relatively short-distance journey to work or commercial journeys between parts of London, given the paucity of internal orbital routes.

25. Relating these functions to both the overall framework for orbital movement in the south-east established 20 years ago in the Strategic Plan, and to the Government's latest

proposals, which have a considerable degree of similarity
between them, we note that:-
-    The "London Bypass" role could in future be taken, at
     least in part, by a combination of the outer east-west
     route together with a lower Thames Crossing route and a
     link between the M3 and M40 (the need for a southern
     east-west route seems much less certain).
-    The second function is that most appropriately met by
     the M25.
-    The third function was intended to be met by an orbital
     route inside London, the GLDP's Ringway 2. This will
     exist to a good standard by the mid 1990s, when the
     planned improvements to the North Circular Road, from
     the M4 at Brentford (near Heathrow Airport) in a
     northern arc picking up all the main national radials,
     and via the East London River Crossing (contro-
     versially) to the A2 in the South-East, are complete.
     No southern route will however exist; as has been shown
     this has always been the most difficult to determine,
     and the Section 5 below refers to the latest proposals.
   26.   It is apparent from the Department of Transport's
response to the Public Accounts Committee that these various
roles, and their traffic-generating implications, have been
poorly understood. Moreover there is a lack of lateral
thinking on the effects of the M25 on the way London city-
region operates in response to the new pattern of
accessibility; on the changes it makes to the operating
conditions and restructuring of economic activity (e.g. in
distribution), or the way it extends labour market catchments.
Looking at it in these terms, we begin to appreciate the
traffic-generating capacity of the London Orbital. This is a
matter for regional planning, which is the theme of the rest
of this paper.

THE M25 AND DEVELOPMENT IN THE LONDON REGION
   27. The relationship between major road investment and
regional development has for long been controversial in this
country. It has taken a low profile in transport circles,
since the Government's Advisory Committee on Trunk Road
Assessment found no causal link in a 1978 report. Empirical
investigations, however, have usually been restricted to
peripheral regions, where road schemes have been built to
stimulate development with disappointing results. The
situation in the London region is quite different. This is a
congested core region, where good accessibility is a vital
criterion of most locational decisions taken by businesses
expanding, restructuring or entering the area. Development
has always followed major transport infrastructure in this
region; it has moulded the way London has expanded. It is of
critical importance not only for the distribution of goods but
also for labour supply and business contact. It is therefore
a reasonable hypothesis that the capacity for movement
realised by the M25, (the largest since the construction of

the London Underground), and the alteration to the pattern of
accessibility around the London region, will have a dominant
influence on the location of development, and on how
development should be planned.

28. This was not taken into account in the piecemeal
planning of the M25 during the 1970s. By 1980, however, when
significant sections of the route were materialising, the
property industry started to realise this developmental
significance and the planning authorities in the "M25 ring"
saw the need for it to be addressed in regional terms.
Ability to do this rests with the London and South East
Regional Planning Conference (SERPLAN), the local government
advisory body for the purpose. This undertook an examination
of the anticipated impact of the M25 in 1981/2 (ref. 13),
stimulated by an awareness that: "by creating the best
accessibility to be found anywhere on the British side of the
Channel, through its orbital linking of the main radial routes
from London to the rest of Britain and the continent, the M25
would create the most attractive locations for future
development. However it would do this where planning policy
is at its most restrictive (the Metropolitan Green Belt), and
at a time when policy is seeking to stem the decline in
London's economy".

29. The SERPLAN report examined:
- the effect on the forces of dispersal from London;
- the strengths and weaknesses of the controls on
  development around the ring;
- the degree of provision for development;
- the changing pattern of accessibility, travel flows and
  travel times.

It found that the M25 would, without policy intervention to
restrict it, reinforce the outward movement of people, jobs
and investment from London. The study concluded, from an
evaluation of relative development pressures, by types of
activity and by geographical sector, that in the western
sectors of the ring, between the A3 and A1 radials, existing
development pressures would be exacerbated and existing
provision would be inadequate, while this is where the Green
Belt is at its most fragmented and vulnerable. In contrast,
the sectors to the east of London have been experiencing
economic decline, but offer major opportunities for
regenerating development. Their poor accessibility hitherto,
being on the "wrong side" of London, is being transformed by
the advent of the M25. SERPLAN pointed out an increasing
regional imbalance, between a prosperous "west" attractive to
investment, and a depressed "eastern corridor" with, in the
early 1980s, little redevelopment interest.

30. Policy proposals were put forward arising from these
conclusions:
- to "steer" development around from the west to the east
  using the new regional accessibility the east will gain
  from the M25 as a promotional mechanism;
- to spread the M25 benefits into London by advancing

strategic transport links to Docklands;
- to promote well-located centres in outer London and
  beyond the M25, able to cope with development pressure.
Having established this framework, restraint policies could be
reinforced in sensitive areas, particularly in the western
sector of the ring.

31. SERPLAN succeeded in persuading the Government to
respond to its work on the matter, in the form of a statement
of regional planning guidance by the Secretary of State for
the Environment in March 1984 (ref. 14). This accepts that
there is an east/west imbalance in the region and that the M25
will assist in steering development to the east. It also
urges planning authorities to provide adequately for the
growth sectors and distribution activities which will require
well-located sites, by implication related to the M25,
although no change to Green Belt policy is seen as justified.
The effect of this statement was wider; the Department of the
Environment was at that time besieged by controversy
surrounding housing land and the ground was fertile for
SERPLAN to suggest that a more comprehensive enlargement of
regional planning guidance should be undertaken, to which the
Secretary of State assented later in 1984. The M25 can
therefore be seen as the catalyst for resurrecting Government
interest in regional planning in the south-east, from its
virtual abandonment following the 1979 election.

32. The further work begun by SERPLAN in 1984 had its
outcome in a report to the Secretary of State in Autumn 1985
advising him as to what further regional planning guidance was
required (ref. 15), to which he responded in the summer of
1986. This confirmed and extended the importance of planning
policy addressing the need to stimulate development on the
east side of London, and clarified the attitude to development
pressure elsewhere around the Green Belt ring through which
the M25 largely runs. This pressure manifested itself,
particularly during the late 1980s, in the form of large out-
of-town retail and leisure schemes (often combined), based
upon the regional accessibility conferred by the M25. While
leisure-only developments have been accepted as compatible
with the Green Belt, most retail or combined proposals have
been rejected, sometimes (in the western sector) for reasons
which include adverse traffic-generating effects in this
already congested area. The exceptions have been in the
eastern sector, where two very large schemes have been
accepted as consonant with the strategy to redress the
east/west imbalance, given their location in the "East Thames
Corridor" identified by SERPLAN.

33. SERPLAN is currently completing a fresh review of
regional planning strategy with a view to requesting the
Government to augment and roll forward the 1986 guidance,
which it is expected to do early in 1991. This takes account
of new concerns over the extent to which large parts of the
region are now exhibiting developmental "overheating", which
points up the need for the remaining areas, which continuing

capacity for further growth, to receive greater priority for infrastructure investment so as to realise that growth. The relationship between transport infrastructure and development in the London region is being even more closely explored in this review. This is being followed through in three situations related to the latest expressions of road policy from the Department of Transport, in the 1989 "Roads for Prosperity" White Paper (paragraph 23 above) and for London itself:-

- In the need to curtail development pressures west of London given the lack of capacity for further growth there; this would seek to influence the studies of outer orbital links in the western and southern sectors of the region, where such links would be seen as stimulating unwanted development.
- Conversely, capitalising on the creation of a "tangential" east/west route in the northern part of the region to attract development to that axis, to the extent it has land and labour capacity to sustain it.
- In London, to extend the planned programme of transport infrastructure in East London to realise the development potential of the East Thames Corridor, while restricting new road building elsewhere to modest local improvements which will not add to development pressures.

34. The regional planning activity over the last decade outlined above, and the growing awareness and understanding of the interaction between transport infrastructure and development, and effect of major accessibility change and its influence on the location and restructuring of activities and the movement patterns they generate, contrast starkly with the lack of awareness of the way the M25 has stimulated traffic growth as revealed in the Department of Transport's response to the House of Commons Public Accounts Committee (paragraph 18). At the least, the Department of Transport seemed to be ignorant of the fact that its sister Department was giving close attention to the impact of the M25 and its contribution to the aims of regional planning strategy.

AN INTEGRAL LAND USE AND TRANSPORT APPROACH TO ASPECTS OF ORBITAL MOVEMENT

35. This part of the paper examines further the options for increasing provision for orbital movement which would relieve some of the functions currently being met by the M25. It does so in a way which seeks to relate such provision to the changing structure of land use activity in London and the South East, and the related pattern of development and traffic generation which ensue.

Orbital Movement Outside London

36. It has already been noted (paragraphs 23 & 24) that the Department of Transport is advancing schemes to create a good quality east-west route across the northern part of the

region, and is studying the case for a Lower Thames Crossing
between Essex and Kent. The latest review of regional
planning guidance by SERPLAN is proposing that this should be
built into an integrated land use and transport strategy which
aims to realise the continuing capacity (in labour and land
terms) which the south-east offers for continuing growth.
Outside London this occurs mainly in the northern and eastern
parts of the region. A new outer orbital corridor can
therefore be envisaged, from the M40 to the M1 (Milton Keynes,
Luton), A1 (Bedfordshire/North Herts), M11 (Cambridge,
Stansted), A12 (N.E. Essex, Ipswich) and via a Lower Thames
Crossing to North Kent and Ashford. Such a route would have
both transport and development functions:-

- it would channel non-London, particularly international
  traffic, away from the M25;

- it would foster a development axis, stimulating
  economic activity in centres on this corridor or on the
  radial routes close to it.

Such an integrated approach would, however, argue against
other ideas for improving orbital capacity outside London, in
the west and south of the region, where developmental capacity
for further growth is becoming exhausted.

Stimulating Development: Orbital and Radial Movement in the East Thames Corridor

37. The East Thames Corridor, extending from London
Docklands to North Kent and South Essex, was identified
following the Government's 1986 Regional Planning Guidance as
offering the main development opportunities which, if
realised, would achieve the aim of removing the imbalance
between the eastern and western sides of London. The main
barrier to the attraction of investment is the poor
accessibility suffered by the Corridor. SERPLAN has
undertaken studies to identify the action required: its 1988
report concluded "the Corridor is poised to realise its
development potential ... (it) would then offer a linked chain
of development opportunities of strategic significance ... a
new growth axis in the eastern part of the region" (ref. 16).

38. To overcome the poor accessibility and the adverse
image which is largely a reflection of it, a transport
infrastructure programme is put forward, much of which has
already been accepted in planning terms, but requires
advancement of implementation. This seeks to relate links
outward to the M25 and beyond to better orbital travel, by
both road and rail. The result would be both good regional
and international access to the Corridor, and easy movement
into it from other parts of London, particularly inner London
in need of new employment opportunities. The radial links
include improvements to the M13 and a new "Thames-side
Industrial Route" south of the River, and to the two Thames-
side railways. The orbital links comprise the extension of
the North Circular across the Thames (ELRiC); the enhancement
of the Blackwall Tunnel and the link from the East Cross

Motorway (the implemented "east-cross route" part of Ringway 1) to the M11; and orbital rail improvements including a river crossing at Woolwich. By 1996, there is the prospect of orbital and radial accessibility on the eastern side of London, between the M25 and the Blackwall Tunnel, which would give the Corridor's main development opportunities the best access in London, together with orbital rail services to both North and South London which will bring people living there within greater commuting range of the biggest single employment-generating developments in the region. Further out, schemes in north Kent would open up further opportunities there.

The Problem Sector: West London

39. The Rendel Palmer and Tritton study (paragraph 20) regarded the quadrant between the A23/M23 Croydon-Gatwick radial and the A40/M40 Oxford radial as that most urgently in need of alternative orbital provision to relieve the M25. As the history of the London Ringways has shown, establishing a route in this sector has been especially difficult. The problem is compounded by the location here of Heathrow Airport, the present surface access to which is generally regarded as unacceptable; its efficient functioning, the growth of traffic through the airport and any proposals for further terminal capacity require significant improvements in this regard. In response to these concerns the Department of Transport set up the "Heathrow Access and South West London Quadrant Study" (HASQUAD) at the end of 1988. Its aims are to examine traffic problems on the approaches to Heathrow (other than from the M25) and of orbital movement in outer South-West London, and to propose solutions.

40. So far the Consultants concerned have examined the types of traffic movement and needs in the quadrant, and have identified particular bottlenecks. The most relevant finding is that most trips are of a short distance, local nature. This suggests a lack of justification for major additions to the strategic road network, although it is likely that the findings reflect the lack of a good existing route. The study is now examining the options put forward in the Rendel Palmer and Tritton study for this part of the M25 ring, but it is already clear, from a land-use viewpoint, that any parallel or "collector-distributor" links in this sector will face enormous environmental opposition. This arises from the fragmented, vulnerable and very valuable nature of the Green Belt, where development proposals associated with the M25 are being firmly resisted by the Planning Authorities and the Department of the Environment. This gives rise to the question of whether the demand for movement west of London should be met, or whether logic lies in an option of planned restraint of road traffic.

Orbital Movement Inside London: The Assessment Studies

41. Decisions about provision for orbital movement within

London have, as revealed, been repeatedly shuffled off for
further study. Such study was revived in 1984 when the
Government announced four "Assessment Studies" of broad
corridors: West London, South Circular, East London and South
London. The first three are concerned mainly with orbital
movement. The South Circular and West London studies are of
particular interest, as they concern the lack of adequate
orbital provision between the A4/M4 and A316/M3 radials in the
west and the A20 in the south east, across South-West and
South London. The purpose of the studies was to identify the
best package of transport measures (including public transport
improvements and traffic management measures as well as road
construction) which would promote accessibility, support
economic development and improve the environment in the areas
traversed.

42. The recommended options from the studies were published
at the end of 1989 (ref. 17). They saw the problem as high
traffic volumes and limited road capacity leading to
widespread congestion, affecting the environment of those
living and working in the areas and reducing the efficiency of
longer-distance movement. The amount of longer distance
traffic was however found to be limited, although some demand
is quite clearly suppressed. Increasing demand for orbital
movement, though, is expected to arise in journey-to-work
terms, particularly eastwards to Docklands. The recommended
options included major, largely on-line, creation of further
capacity on the South Circular route between Eltham (A20) and
Wandsworth (A3) including tunnels under Clapham Common and
Forest Hill, a new route largely in tunnel from Chiswick
(A4/A316 leading to M3) through Barnes to Wandsworth, and a
new relief road from Hammersmith (A4) to Wandsworth (the
"western environmental improvement road", in essence a
throwback to the old west cross route part of Ringway 1).
Improvements to public transport links, involving extensions
of the London underground network, were proposed for inner
South London.

43. Following consultation during the winter of 1989/90,
the Secretary of State for Transport has rejected the major
road construction schemes recommended, in favour of an
approach based on public transport (with several schemes
endorsed for "detailed attention"); junction improvements and
local widening only to the roads, together with traffic
management and parking control measures (ref. 18). As well as
overwhelming opposition to the major road options, the reason
for this stance seems to be the lack of real demand for longer
orbital movement across South London and that such schemes
would simply not be value for money. In part, the lack of
demand or support for an improved southern orbital within
London relates to the relatively poor development of
employment or commercial centres in South London, compared
with the situation around the North Circular. While this is a
reflection of relatively poor accessibility, it is not part of
London's planning strategy to foster growth points on this

corridor. The outcome, therefore, seems to put the final nail in the coffin of a southern equivalent to the North Circular, and point South London's future towards public transport, the main orbital requirement being based on where new employment is being created.

The Public Transport Alternative: Croydon Light Rail

44. In outer South London, little attention has been given to provision for orbital movement since the abandonment of Ringway 3 in the early 1970s, since when it has been accepted that a road solution is totally unacceptable on all counts. Recently, however, one of the most interesting proposals for new orbital movement in London has emerged, which stems directly from the development situation and planning strategy for the area. This is the scheme for a light rail network centred on Croydon. Croydon, in the outer southern part of the London conurbation, is the largest office centre outside Central London and is one of three primary retail centres in South London, with an extensive catchment of great strategic importance. Its economic health and growth is suffering from acute traffic congestion which, unless resolved, is likely to lead to activities relocating elsewhere.

45. A study of the situation has recently reported (ref. 19), from which a light rail solution is favoured by all parties, showing significantly better value for money than other options. Opposition to new road building was confirmed, and complementary measures to restrain car use have been endorsed. The initial network is proposed to comprise three lines, centred on Croydon's mainline British Rail station: to Wimbledon in the west, Beckenham in the east and the suburb of New Addington to the south east; further extensions are envisaged. The benefits would significantly improve Croydon's image and economic development prospects as a key element in London's development structure, particularly in extending its labour catchment east and west, and providing a stimulus to investment in both the office and retail sectors. The overall conclusion, accepted by the Department of Transport, is that the scheme is financially viable, and is capable of realisation over a five year period.

CONCLUSIONS

The Lessons of the M25

46. The history of the planning and implementation of the M25 has shown how the London region has come to rely on one orbital motorway, from a "grand design" formulated 45 years ago in Abercrombie's Greater London Plan and repeated in the late 1960s by the Greater London Council, for four orbital routes within, around and outside the conurbation, each with complementary functions. With the benefit of hindsight, three points can be made in summary:-

- the route built, largely through the Green Belt, was probably the only one which was financially feasible and environmentally acceptable;

- the route developed incrementally and not as part of an
  integrated strategy, despite the 1970 Regional Plan
  which saw the need for complementary outer tangential
  routes related to the future pattern of regional
  growth;

- the outcome is a situation where the mix of functions
  the M25 is required to sustain (reflected in the
  pattern of intersections) has led inevitably to serious
  overloads and congestion.

47. We now have the "M25 problem" with traffic levels well
beyond forecasts and likely to be extremely costly to resolve
and in part intractable, particularly in the west. This seems
to be due to a failure to recognise how a major increase in
accessibility on a situation of suppressed demand for
movement, would stimulate the restructuring of economic and
social activity in a dynamically interactive way. The story
reveals, however, that with better planning mechanisms the
problem should have been anticipated and catered for.

Towards an Integrated Strategy

48. The conclusion is therefore that in a congested core
region such as London, major transport investment and
development strategy need to be much more closely associated.
Such association would shift the conception of such new routes
from providing for growth of existing traffic patterns to
their role in providing for metropolitan growth and change.
The M25 shows that in a congested region, the implications of
creating a quantum increase in accessibility and enabling new
patterns of movement have not been understood, when economic
activity (whether services or manufactures) puts a great
premium on such accessibility, for labour as well as inputs
and outputs. Activities will be very attracted to that
accessibility, and will restructure dynamically, as
exemplified in the distribution, retailing and leisure
industries. The M25 has had a demonstrable effect in
extending labour markets, generating new journey to work
traffic, and has stimulated development in areas not
previously viable.

49. Latterly, regional planning has pointed up the close
relationship between such transport infrastructure and the
future development of the region, related to its geographical
capacity for growth (ref. 20). In this regard the imbalance
between an increasingly overheated west side and an east with
continuing ability to sustain development indicates that this
growth should be stimulated by new infrastructure: in the East
Thames Corridor, and in the outer north and east of the
region. The eastern area would then become more accessible
and attractive to growing and restructuring activities. The
west side of London, conversely, faces a critical decision:
the Rendel Palmer and Tritton study concluded that the western
sectors of the M25 reveal strong continuing potential for
traffic growth, unless restrained by capacity limitations.

The time has come to consider such restraint, particularly as further capacity for movement here would add to development overheating and environmental disbenefits.

The Future of Orbital Movement around London

50. An approach which relates the functions and requirements of orbital movement to the proposed development strategy for London and the south-east would have the following elements, close to that indicated in the National Audit Office's 1988 report:-

-   Diverting "London Bypass", including international, traffic onto outer tangential routes on a northern and eastern axis related to the new regional development strategy, from Milton Keynes to the A12 and then North Kent.
-   Evolve London's internal urban distributor network in relation to its changing development structure; this will include in road terms the North Circular from the M4 in the west linking all the long-distance national radials across North London to the "corridor to the continent" in the south-east, together with links in the East Thames Corridor, but will otherwise have a public transport emphasis, including such orbital routes as an improved North London line, extensions to the rail network in South London, and the Croydon Light rail network. Together these would make East London the most accessible part of the metropolis.
-   This leaves the M25 to function as the city-regional motorway channelling London-bound traffic around its outer edge to the most appropriate access routes to destinations.

The M25 will become dual 4-lane, but we shall not see further on- or near-line improvement. Instead, there will be "planned restraint" of traffic in the western sector, alongside improved provision of public transport as proposed at Croydon. This will take place within an overall strategy of curtailing the growth of both traffic and development to the capacity of the area to sustain it, in a way compatible with the environment.

REFERENCES
1. ABERCROMBIE PROF. PATRICK. Greater London Plan 1944. London, HMSO 1945.
2. GREATER LONDON COUNCIL. Greater London Development Plan Statement. London, GLC, 1969.
3. GREATER LONDON COUNCIL. Greater London Development Plan Report of Studies. London, GLC, 1968.
4. SOUTH EAST JOINT PLANNING TEAM. Strategic Plan for the South East. London, HMSO, 1970.
5. DEPARTMENT OF THE ENVIRONMENT. Report of the Panel of Inquiry into the Greater London Development Plan, vol. 1 report. London, HMSO, 1973.

6.   GREATER LONDON COUNCIL. Greater London Development Plan including Notice of Approval by the Secretary of State for the Environment. London, GLC, 1976.

7.   SOUTH EAST ECONOMIC PLANNING COUNCIL. Background Paper to the Modified Greater London Development Plan (SEEPC (76) 16), 1976.

8.   DEPARTMENT OF TRANSPORT. Policy for Roads in England, 1983. London, HMSO, 1983.

9.   CONTROLLER AND AUDITOR GENERAL, National Audit Office. Road Planning. London, HMSO, 1988.

10. HOUSE OF COMMONS COMMITTEE OF PUBLIC ACCOUNTS. SESSION 1988/89. Road Planning. London, HMSO, 1989.

11. DEPARTMENT OF TRANSPORT. M25 Review. Prepared by Rendel, Palmer and Tritton Consulting Engineers. London, HMSO, 1989.

12. DEPARTMENT OF TRANSPORT. Roads for Prosperity, Cm. 693. London, HMSO, 1989.

13. LONDON AND SOUTH EAST REGIONAL PLANNING CONFERENCE. The Impact of the M25, SC 1706. SERPLAN, 1982.

14. DEPARTMENT OF THE ENVIRONMENT. Strategic Planning Guidance for the South East, Planning Policy Guidance Note 9. London, HMSO, 1988.

15. LONDON AND SOUTH EAST REGIONAL PLANNING CONFERENCE. South East England in the 1990s: a Regional Statement, RPC 450. SERPLAN, 1985.

16. LONDON AND SOUTH EAST REGIONAL PLANNING CONFERENCE. Increased Activity in the East Thames Corridor, RPC 1000. London, SERPLAN, 1988.

17. TRAVERS MORGAN. South Circular Assessment Study. SIR WILLIAM HALCROW & PARTNERS. West London Assessment Study. Summaries of Reports prepared for the Department of Transport. London, 1989.

18. DEPARTMENT OF TRANSPORT. Parliamentary Statement by Cecil Parkinson, Secretary of State, and Press Notice no. 96. London, 27/3/90.

19. THE MVA CONSULTANCY. Croydon Area Light Rail Study. A report to the London Borough of Croydon. London Transport and British Rail, 1990.

20. LONDON AND SOUTH EAST REGIONAL PLANNING CONFERENCE. Shaping the South East Planning Strategy. Consultation Paper, RPC 1660, April 1990.

THE M25 LONDON ORBITAL MOTORWAY

Fact Sheet

General Information

Length: 119 miles (188 km)

Radius: Generally 35 kms but between 13 and 22 miles (21-35 km) from Charring Cross
13 miles (21 kms) from Charring Cross at Junction 24 (Potters Bar)
22 miles (35 kms) from Charring Cross at Junctions 5 and 28 (Sevenoaks and Brentwood).

Junctions: Number - 32 (inc 21 and 21A)
Type - Grade separated Interchange and Junctions

Structures: Number - 234 bridges on M25
3 tunnels

Landscaping: 2.1 million trees and shrubs

Emergency Services: 452 telephones linked to 4 police centres

Public Services: 3 Service Areas

South Mimms - Herts   Westerham - Kent   Thurrock - Essex

(Department policy is to try to provide one every 30 miles)

Cost: £1,000 million without allowing for inflation
£5.8 million came from European Community

Legal Procedures: 39 separate Public Inquiries

Population: The population within the M25 circle has not been calculated, however Greater London forms a substantial part of the population and it is at present recorded as 6,735,000 people.

The M25 Motorway probably encircles 7-8 million people. Please see Appendix A.

Standard

Dual 3 lane Motorway

Exceptions:- Junction 11 to Junction 13 (being widened to Dual 4 lane)

Junction 13 to Junction 15 (Dual 4 lane)

Climbing Lanes:- There are two existing climbing lanes:

(i) Swanley (Junction 3 to 2). An unsignposted climbing lane in an anti-clockwise direction for 500 metres.

(ii) Upshire (Junction 26 to 27). This is signposted as a crawler lane and is 1.48 kilometeres in length.

ORBITAL MOTORWAYS

Traffic

For annual average daily traffic details (please see Appendix B)

Interchanges

Spacing:

JUNCTION NUMBERS	DISTANCE	
	MILES	KILOMETRES
1-2	1.6	2.5
2-3	3.3	5.3
3-5	8.1	12.9
5-6	4.8	7.8
6-8	4.5	7.2
8-9	4.5	7.2
9-10	4.3	6.9
10-11	5.9	9.4
11-12	2.4	3.8
12-13	2.5	4.0
13-14	0.9	1.5
14-15	2.0	3.2
15-16	4.3	6.8
16-17	4.8	7.7
17-19	6.0	9.6
19-20	3.6	5.8
20-21A	1.9	3.0
21A-22	4.2	6.7
22-23	3.8	6.1
23-24	2.7	4.3
24-25	5.3	8.5
25-26	3.8	6.1
26-27	4.1	6.6
27-28	8.1	12.9
28-29	5.4	8.6
29-1	6.1	9.6

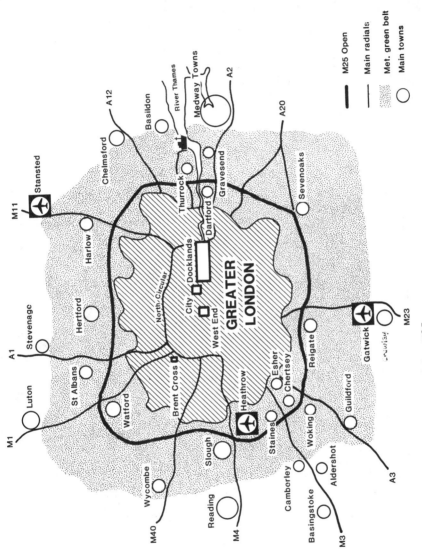

APPENDIX A. Area encircled by M25

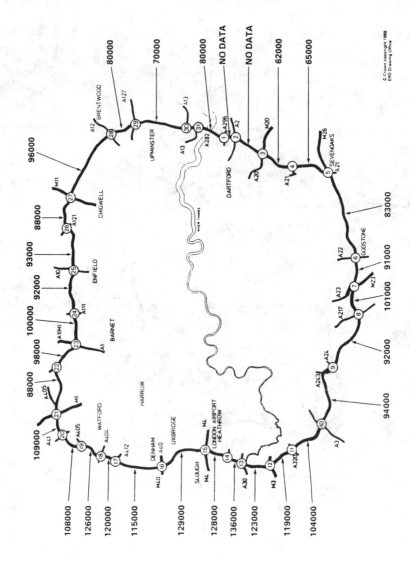

APPENDIX B. Department of Transport M25 AADT 1988

Discussion

W. MCALONAN, Consultant (Babtie Shaw & Morton), Glasgow
I would like to make the point that almost 70% of the control action for CITRAC is concerned with pre-planned closures and diversions associated with maintenance. I also hope that a dynamic system of automatic incident detection is not seen as a substitute for closed circuit television. The advantages of visual surveillance for the purpose of monitoring action taken during major events and a general safety overview.

PROFESSOR P.J. HILLS, University of Newcastle-upon-Tyne
1.What reason do you give for the lack of ramp metering facilities on more recent motorways (notably M25 and M40) in the light of the successful experiments on the M6 which you have described?
2.Do you see a design implication for on-ramps from the M6 experiments? The site chosen in the M6 was in fact one with on-ramps designed in a pre-congestion era. Should not on-ramps now be designed with 'reservoir' capacity built in, lane dividers unconventional (e.g. 'wig-wag') signals, and vehicle detection equipment installed on gantries or at the roadside?

PROFESSOR P. HALL, University of California, Berkeley
In the USA, ramp metering has been extensively associated with priority for HOVs (high occupancy vehicles). Why hasn't a similar opportunity been seized in the British examples?

J.M.H. KELLY, Rendel Palmer & Tritton, London
I feel that the general case for ramp metering is understood and accepted, but would like to point out that where there is a lane gain after the junction, such as each of junctions 10 to 14 on the D4 section of the M25, ramp metering seems inappropriate. In my opinion, the optimum solution in such cases would be to achieve access control by reducing the number

of entry lanes at the end of the entry slip to one
(by coning in a taper), so as to match the capacity
of the extra lane downstream of the junction.
French engineers are using this form of access
control with success. It can be implemented quickly
and cheaply, and removed equally quickly if
necessary.

A.D.R. TETLEY, Frank Graham Group, Reading
I would like to have your opinion on the use of
charges or tolls to optimise the scale and usage of
new roads.

G.H. FRENCH, Scott Wilson Kirkpatrick & Partners, Basingstoke
As my firm is responsible for the HASQUAD (Heathrow
and South West Quadrant Orbital Movement) study
mentioned by Mr Simmons, I'd like to clarify one
point.

It is true that phase 1 of the HASQUAD study
determined that some 'short distance movements' used
the M25. However, it also identified the problem
that there was very little information on the nature
of the existing travel patterns. The current phase
of the HASQUAD study has therefore carried out
surveys on the M25 on- and off-ramps in the south
west quadrant to collect this information.

G. JONES, G. Maunsell & Partners
There have been several references to the dual roles
of orbital routes and to unfocused movements: there
is clearly a lack of information on trips using the
motorways and the relative magnitude of 'short
distance' and 'long distance'.

The problem of obtaining accurate information on
trips using motorways, particularly the long
distance motorway to motorway movements, has been a
major problem for some years. It has been
impossible to carry out the traditional
origin-destination interview surveys on motorways
for obvious reasons.

Maunsell has developed a new technique in response
to the need to obtain trip data on the orbital
motorway around the north west of Manchester.
Partial origin-destination information is obtained
using video cameras to record the registration
numbers of vehicles at carefully chosen locations,
enabling accurate estimates of motorway through
traffic. In this way, information on trip patterns
within a complex motorway network such as that
surrounding Greater Manchester can easily be
obtained.

The technique essentially consists of recording
information on all vehicles on the motorway using

high quality video cameras positioned on overbridges. From these readings, vehicle registration numbers are transcribed together with type of vehicle, time of day, and other information if required. The data are then matched to determine the volume of traffic between the chosen locations. Using this technique much higher sample rates can be achieved than with manual field recording methods, giving higher confidence levels and less variable results.

Studies carried out for the Department of Transport into the proposed Greater Manchester Western and Northern Relief Road have used the technique to determine the volume of traffic between:

(a) M62, M61 and M66 to the west and north of Manchester
(b) M56/M6 and M62/M602 to the west of Manchester
(c) M63/M56/M56 airport spur and the M6/M62 to the west of Manchester.

M.A. SELFE, Essex County Council, Chelmsford

While supporting the provision of a new outer transport corridor to the east of London, local representatives are becoming increasingly concerned at the attendent development which follows and diminishes the accessibility gains.

M. SULLIVAN, Council for the Protection of Rural England

If the M25 had opened fully in 1986 with full ramp metering, reservoir space for queuing, metering from other motorways (as exists in two US cities), incident detection and gantry signs, would we be worried now? It could have held flow down to free flow at 55 mile/h or so.

I suggest that to do this now would hold down the commuter use of the orbital while maintaining its function as a through route and bypass for London. If 2000 vehicles/lane could be maintained at 50-55 mile/h, 12000 vehicles/hour could be carried in a managed way and the case for further lanes evaluated on that basis.

J.A.L. DAWSON, Scottish Office, Edinburgh

You have said that the forecasting process on M25 underestimated flows because it didn't consider induced development. I have been told that on the newly opened Edinburgh bypass flows are generally quite close to those forecast if correction is made for changes in the nationally issued road traffic forecasts. Regarding the M25, how do you assess the relative importance of errors made in underestimating

(a) national economic and hence traffic growth, and
(b) route attractiveness?

J. WOOTTON

I should like to reply to the comments made by
Professor Hall, Professor Hills and Mr Dawson.
First, I can give no reason for the lack of ramp
metering facilities on the more recent motorways, in
particular the M25, and I believe that the question
is better directed to central Government for an
answer. The first ramp metering site at junction 10
on the M6 became operational in the spring of 1986.
It continues to be successful, as are the other
sites that have been opened on the M6, and it seems
surprising that the methods have not been adopted on
other motorways where benefits would undoubtedly
accrue.

I agree with Professor Hills that the design
criteria for the on ramps at motorway intersections
should recognise that ramp metering and other forms
of control may be required at some later date. The
experience gained in implementing the M6 schemes
will help to establish the appropriate changes to
the design criteria.

Professor Hall's point concerning the association
of ramp metering with priority for high occupancy
vehicles is recognised. At some future date the two
may be linked, but the cautious view that was, and
still is, adopted towards ramp metering has meant
that resistance to elementary control, i.e.
signalling, has had to be overcome before other
priorities can be considered.

Transport planners have always known that the
amount of traffic, both in terms of the number of
vehicles and their use, is strongly related to the
state of the economy. In periods of strong economic
growth, there is strong traffic growth. The most
recent national road traffic forecasts have
recognised, after the event, that the economic
growth during the 1980s has been much stronger than
had been anticipated at the beginning of the decade.
As a consequence of the higher economic growth, our
traffic forecasts, including those for the M25, have
probably been underestimated by around 10%.

Mr Dawson is also concerned about the
attractiveness of the M25 both as a route and for
the relocation of population and employment. It is
sad that there are no accurate figures on the
relocation of population and employment that has
resulted from the construction of the M25, since, as
I have shown, this is perhaps a very important
effect. An extrapolation of the observations I made
in my paper suggest that perhaps 10% to 25% of the

traffic currently using the M25 is the result of relocated population and employment. I must emphasise that the assessment procedures will not have taken these effects into account, although in principle there are methods of evaluating them.

M. SIMMONS

In response to Mr Jones, I would only say that the more information that can be obtained on the different types of movement using the M25, to quantify the extent to which it is performing different functions, the better.

In response to Mr Selfe, the first point I would make is that in a region like South-East England development will inevitably follow the creation of such new strategic transport capacity; the question is whether integrated planning should seek to capitalise on it (with implications for the new route's traffic capacity and pattern of intersections) by locating new development in relation to it, or should regard such a new route solely for through movement and not provide access to it. My second point is a land use one: in putting forward the proposition of a new development corridor on the northern and eastern flanks of the South-East, SERPLAN is not visualising development all along it, but that certain growth points should be recognised, some of which already exist, such as Milton Keynes, Northampton, Cambridge, Harwich/Ipswich and in Kent, and that development should be provided at those general locations.

In reply to Mr French, my point about the HASQUAD Phase 1 findings was more about the nature of trips on roads inside the M25 than on the motorway itself, where I agree there is a need for more information and I am glad the current HASQUAD work is addressing it. I was arguing that, from Phase 1, there is little orbital traffic justification for adding to the strategic road network inside the M25, to set against the great environmental opposition.

Orbital motorways in the region of Paris

J. G. KOENIG, Division des Infrastructures et des Transports,
Direction Régionale de l'Equipment d'Ile de France

SYNOPSIS

Two major orbital motorways are being
implemented in the region Ile de France : the A.86
and the "Francilienne". These projects have an
essential function in the regional road network,
but rise important difficulties. They received a
strong support from both the government and the
regional council, so that their completion might be
expected in about 8 to 10 years.

– GENERAL BACKGROUND

1-1 The region Ile de France

The region of Paris –now called "Ile de
France"– covers about 12 000 km2.

Its population, actually 10.2 millions
inhabitants, has been growing until the early
eighties, but is now roughly stabilized. Population
has decreased in central areas, and is increasing
in outskirts and new towns.

In 1965, forecasts made for the long term
plan were expecting a population of 14 millions
inhabitants for the year 2 000. This figure has
since been reduced to about 11 millions.

1-2 Political organization (fig. 1)

Beside the traditional local authorities
(communes and departements), the regional council
emerged recently as a major partner of the
government.

1-2-1 Communes

There are about 1 300 communes (cities or
villages) in the region, including small rural
villages as well as the powerful city of Paris (2
millions inhabitants).

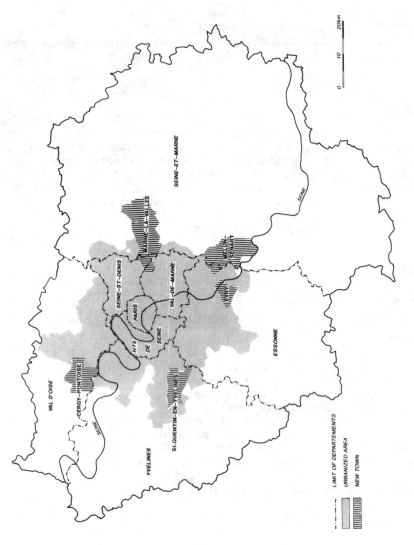

Fig.1 – THE REGION ILE DE FRANCE

1-2-2'Departements
There are 7 "departements" in the region ; the city of Paris is considered as the 8th departement.

Each departement has an elected council (conseil général) and has -like the communes- its own road network, some of these roads being important.

1-2-3 Regional council
In march 1982, a law instituted the regional councils as a new level of local authority, the regions covering several departements.

Each region has an elected regional council and deals with such matters as roads, public transport, education, regional planning and economic development.

The regional council has no road network of its own, and can only subsidize the roads belonging to the state, departements and communes.

In 1989, the regional council of Ile de France financed 1,640 millions Francs (i-e.160 millions £) [1] for investments on state roads, while the state itself invested only 600 millions Francs on its own road network. Thus the regional council is obtaining a growing share of the decision power concerning the national network.

1-3 Regional planning and road planning

1-3-1 Long term regional planification
The general long term regional plan in Ile de France is the "SDAURIF"[2]. This document, established by the state after consultation of local authorities, describes the general land use guidelines and includes the long term road and public transport maps.

Land use plans are established by the communes and must be compatible with the regional plan.

1-3-2 Regional road plan
The regional road plan, as a part of the general plan, has to comply with the functional necessity of a regional coherence in the long term network, and with the political necessity of some

---

(1) 1 £ = 10 francs
(2) Schéma Directeur d'Aménagement et d'Urbanisme de la Region Ile de France

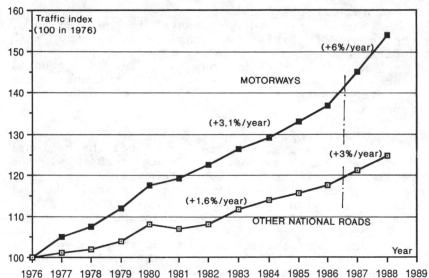

Fig.2— EVOLUTION OF TRAFFIC ON NATIONAL ROADS IN THE REGION ILE DE FRANCE

local acceptation of the land reservations for future roads. Mayors of communes where a future road is planned have a tendancy to contest the road, or to accept it, provided it is transfered to another commune.

Thus the balance between coherence and acceptability is sometimes difficult to find. Since the first "SDAURIF" was established, several cancellations have been brought in the road plan. They have been justified by the reduction of the expected long term population, and by the priority given to public transport for radial trips. These reductions concerned mainly radial roads ; ring roads have generally been preserved.

1-4 Actual tendancies

1-4-1 Road traffic (fig. 2)

Albeit the overall population has been roughly stable since 1980, the road traffic has been increasing at a rate of about 3 % per year on national roads (5 % on highways).

This is due to the increase of car ownership and of individual mobility, and also to the transfer of population from central areas to the outer zones of the urbanized area, where public transport is poor and car use high. These reasons will not disappear in the coming years.

84

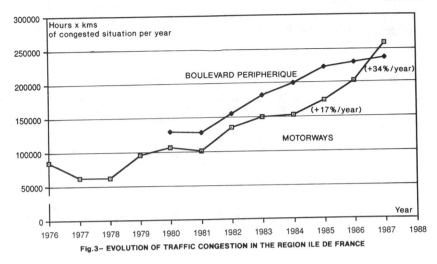

Fig.3- EVOLUTION OF TRAFFIC CONGESTION IN THE REGION ILE DE FRANCE

1-4-2 Traffic congestion (fig.3)

The traffic increase, associated with a long period of low investment, resulted in a very fast increase of traffic congestion. Measured in terms of hours x kms of congested situation, the congestion on highways and expressways has been increasing since 1975 at an explosive rate of 17 % per year.

The yearly economic loss in wasted time is estimated at 5 billions Francs in 1988.

1-4-3 Road investments (fig.4 and 5)

Expressed in constant Francs of 1988, the yearly investment in the region for national highways and roads has fallen from 4.2 billions Francs in 1971 to 1.2 in 1986.

This fall resulted in such an increase of congestion that the government and the regional council decided in 1987 and 1989 an important program of new investments for the next years.

Two sources of finance will be used :

(i)   Budget resources

The contract negociated in 1989 between the government and the regional council for the next five year period (1989-1993) includes a commitment from both partners to finance together 11 billions Francs for national roads, a substantial increase on the 8 billions of the last 5 year period.

An extension of 1.5 billions Francs of this program was decided in october 1989, for the A 86 ring road.

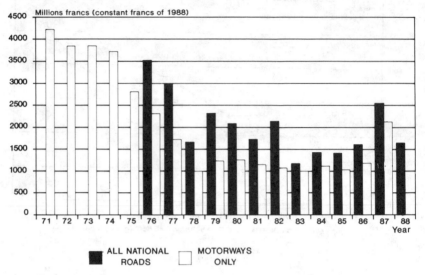

Fig.4– EVOLUTION OF INVESTMENTS FOR NATIONAL ROADS IN THE REGION ILE DE FRANCE

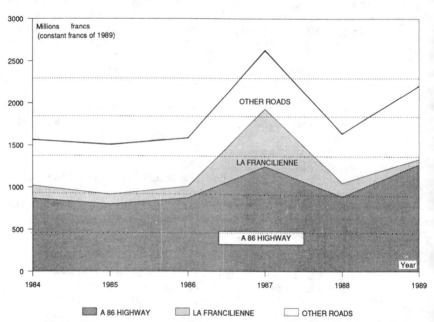

Fig.5– THE IMPORTANCE OF ORBITAL MOTORWAYS IN THE INVESTMENTS
FOR NATIONAL ROADS IN THE REGION ILE DE FRANCE

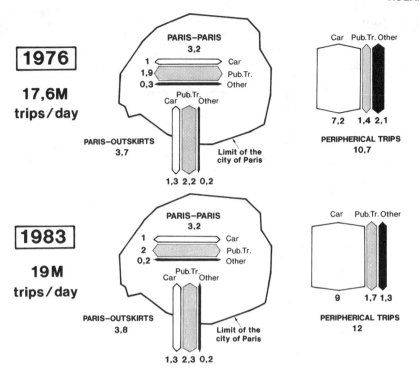

Fig.6— MOTORIZED TRIPS IN THE REGION ILE DE FRANCE
for each mode and geographical type

(millions of daily trips,in 1976 and 1983)

(ii) Toll highways
     In april 1987, the government decided that 5 highways included in the long term plans would be conceded to private or public operators, thus becoming toll highways.
     A sixth concession (on A 86 ring road) was added recently.

     From 1990, the yearly road investments for national motorways are therefore expected to reach 2.6 billions Francs per year, or 3.5 billions Francs if including the toll highways.

3— IMPORTANCE AND LOCATION OF RING ROADS IN THE REGION ILE DE FRANCE

3-1 An important and increasing potential of traffic (fig.6)

     The importance of ring roads is illustrated

by the general home interview surveys of 1976 and 1983 (the next one is expected in 1991).

In 1983, 19 millions trips were made on a daily basis in the region Ile-de-France (17,6 millions in 1976). A majority of them (63 %) were peripherical trips, whose origin and destination were not in the city of Paris.

31 % in 1983 of all motorized trips were made by means of public transport (the proportion was the same for 1976). This proportion was an average figure between 61 % for central trips with origin and/or destination being in the city of Paris, and only 14 % for peripherical trips. At peak hours, it was increasing to 39 % of all trips in 1983.

Car trips represented in 1983, 60 % of all motorized trips (54 % in 1976), but 34 % of central trips, and 75 % of peripherical trips. At peak hours, the car use ratio was only 52 %.

Other means of transport (mainly 2-wheeled) declined considerably from 15 % in 1976 to 9 % in 1983.

When considering car trips in particular, the surveys reveal that :
(i)   the number of central trips stayed constant during the last 10 years ;
(ii)   80 % of all car trips are peripherical trip ;
(iii) these peripherical trips are rapidly growing, at a rate close to 3 % per year.

This tendancy will probably persist, as it expresses the transfer of population from central to peripherical zones, and the high use of car in such peripherical zones as mentioned earlier.

As a result the highest priority in road programming consists in implementing ring roads, in order to allow peripherical car trips (80 % of all car trips) not to pass through central areas and increase congestion on radial roads.

3-2 The 3 levels of ring roads (fig.7)

Three ring roads are implemented, or planned in the long term :
(i)   **The "Boulevard Périphérique" of Paris** was built between 1960 to 1973 as a belt for the city of Paris. It isn't a national road, but a communal road of the city of Paris. It provides a standard of dual 3 lane on the southern section and dual 4 lane on other sections.

The "Boulevard Périphérique" is the only complete ring road in existence in the region. It

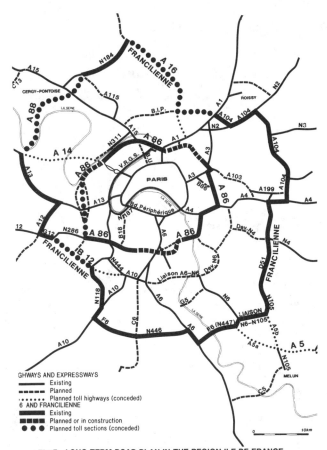

Fig.7– LONG TERM ROAD PLAN IN THE REGION ILE DE FRANCE
A86 AND FRANCILIENNE ORBITAL MOTORWAYS

is congested 6 to 10 hours a day. 1 million
vehicles use it everyday, and the daily traffic is
about 220 000 vh. on dual 4 lane sections, and
180.000 vh. on dual 3 lane sections.

(ii) The **A 86 highway** is the major ring road of
the state network. It's by far the most important
road project in the region, and represents more
than half of the total road budget for national
roads in the region of Paris.

(iii) **The "Francilienne"** is the third level of ring
road, with the largest diameter. It corresponds
with a new concept of an outer ring road introduced
in the long term plans in 1984.

The situation of these two ring roads will
now be described with more details.

4– THE A 86 HIGHWAY (Peripherique d'Ile de France)

(fig. 7 and 8)

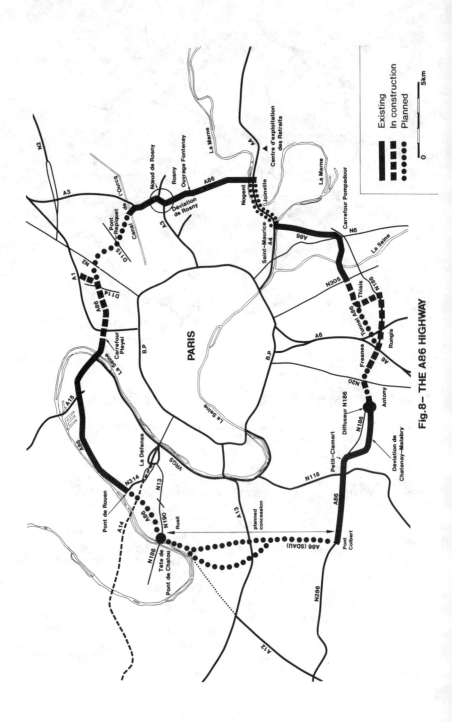

Fig.8– THE A86 HIGHWAY

4-1 Actual situation

The A 86 highway is the main ring road in the long term plan of the region Ile de France.

It is situated at a distance ranging between 3 and 9 kilometers from the Boulevard Périphérique of Paris, mostly in dense urbanized areas. The majority of its length follows the actual national road RN 186, which offers only poor characteristics.

The total lengh of A 86 is 80 kilometers, of which 45 kms are presently in service (including 4 kms of common sections with A 4 and A 3 highways, to be doubled later) ; 7 kms are in construction, and 18 kms are in project.

In its final stage, the A 86 will offer a standard of dual 3 or 4 lane, depending on the considered section, with grade separated junctions. New sections are usually built with dual 2 lane, and are designed to be widened later without further expropriations.

A significant stage will be reached by the end of 1990, when the A 86 will provide a continous link whithout traffic lights in the east of Paris between the A 1 highway (in the north of Paris) and the A 6 highway (in the south).

This link is expected to substantially relieve the traffic congestion on the Boulevard Périphérique and on the actual RN 186 road (parallel to A 86). The opportunity of this relief offered on the RN 186 is used to implement on this road a new dual lane reserved for buses (project TVM : Trans Val de Marne) and a tramway between Saint Denis and Bobigny.

4-2 Costs and development program

Recent sections of A 86 usually cost between 500 and 600 millions Francs per kilometer ; difficult sections, such as the underground crossing of the Marne river, may cost up to 1 billion per kilometer.

This cost is essentially due to the necessity of guaranteeing good environmental conditions in densily urbanized areas.

Two sections are in tunnels (an existing section under the Marne river, and the future toll tunnel in the western section). Several other sections are covered, in order to avoid noise disturbance. Underground solutions are requested by a part of the population in another section in Joinville, where less expensive solutions are considered as sufficient in the project to protect the environment.

According to the recent program contract signed by the government and the regional council, the two partners will together invest each year about 2.2 billions Francs for national roads from 1989 to 1993 (i.e. 11 billions Francs for the 5 years). At least 60 % of these sums are financed by the regional council.

In 1989 and 1990, 50 to 60 % of this investment will be used for the A 86 ; this highway has in fact clearly received the highest priority (fig.5)

Despite these high investments, the expected date for the complete opening of the A 86 still seemed to belong to the third millenary : as a matter of fact, the contract plan could finance 5 billions of the A 86, when the total cost estimated for achieving a complete ring and improving existing sections is estimated at 20 billions (including 7 billions for the western section between Rueil and Pont Colbert).

Last october, the government decided :

(i) to propose a supplemental program of 1,5 billions for the A 86 for the next 4 years ;

(ii) to concede the western section of the A 86 (11 kms, mainly underground for environmental reasons), which will as a consequence be a toll highway with private financing.

Under these conditions a completion of the A 86 could be expected for the year 1998, provided the total financial effort is maintained.

5- **THE FRANCILIENNE RING ROAD** (fig.7)

The Francilienne ring road is a rather recent idea, and all sections are not yet completely defined.

In the first long term plan of 1976, the outer ring road was constituted by several projected highways such as A 87, C5 or C6. Nevertheless, several sections of these projects disappeared in modifications of the long term plan brought between 1980 and 1984 ; at the same time the idea emerged that other separate links of the road plan could be connected in order to offer, at least in the eastern and southern part of the region, a new outer ring road.

In the new long term road plan of 1984, this concept was approved as the "Rocade des Ville Nouvelles", which was linking 4 of the 5 new town of the region (Saint Quentin en Yvelines, Melu

Sénart, Evry and Marne la Vallée) ; it was obtained by connecting several sections of the road plan (A 104, CD 51, F 6).

This ring road could be continued in the north and west of Paris with other existing or projected highways in order to obtain a complete ring road linking the 5th new town (Cergy-Pontoise).

In 1987, a special road program was decided by the government and the regional council to boost the construction of new roads.

In this special program a high priority was given to the new outer ring road, which could offer strong support to the development of new urbanized areas (specially in new towns), at a cost per kilometer 10 times lower than the A 86. The previous "Rocade des Villes Nouvelles" received its new name "la Francilienne".

With the product of a new tax on offices decided by the government, the regional council could finance the complete linkage in the south and east of Paris ; this section was completed, and opened to the traffic at the beginning of 1990.

Finally, the special program of 1987 decided to implement the Francilienne as a complete ring road, also in the west and north of Paris. This goal could be achieved by continuing the east and south sections with some existing sections (A 10, RN 118, A 12, A 13) and with new highways mentioned in the long term road plan (B 12, A 88, A 16, B.I.P.). For this purpose the government decided that these 4 new sections would be financed by concessions and become toll highways. These concessions are still in study and discussion ; construction is not expected to begin before 1991-1992.

In its actual shape, the Francilienne ring road is situated at a distance of 25 to 35 akilometers from the center of Paris. Its total length is about 190 kms, of which :
- 131 kms are in service ;
- 59 kms are in project, for an expected cost of about 4 billions Francs (financed by concessions).

The standard configuration is a dual 2 lane motorway, with grade separated interchange and junctions. A possible widening to dual 3 lane is reserved on most new sections, but would be very expensive (2 billions Francs) on existing sections.

Some particular difficulties should be mentioned :

- common trunks with radial highways (A 10, A 12, A 13, A 16, A 4) may cause traffic congestion by mixing radial with ring road traffic, and should be widened in priority ;
- the Francilienne is situated at the level of the "Green Belt" of Paris, and precise land use plans will be necessary to avoid undesired urbanization

CONCLUSION

The two projects A 86 and Francilienne can illustrate both the necessity and difficulty of implementing ring roads in the region of Paris.

Orbital motorways present very specific characteristics when compared with other motorways, making them politicaly wise more acceptable :

(i) there is a strong logical consistency in completing a ring (if it's not complete, it isn't a ring) ;

(ii) orbital motorways may clearly help relieving traffic congestion in central areas and on radial roads ;

(iii) the competition (or possible substitution) with public transport is lower for ring roads than for radial roads ;

(iv) ring roads can provide a strong support to a strategy of urban development based upon new poles linked together, like the new towns in the region of Paris.

The example of the A 86 and the Francilienne suggests that the "syndrom of ring completion" may have played an important role in the decisions taken for implementing these two orbitals.

The experience of A 86 and the Francilienne also confirms more conventional conclusions. Political and local acceptability requires an important and continuous effort in explaining the role and utility of those ring roads. Preserving the urban and natural environment implies a constant dialog with local populations and the acceptation of important costs (covered sections or tunnels). Financing such expensive infrastructures needs new resources, that can be provided by special taxes or by conceding toll sections to private or public operators. Undesired effects on land use must be controlled.

Nevertheless, it can be hoped that the opening of long continuous sections of A 86 and the Francilienne in 1990 will provide an important improvement of road traffic conditions in the whole region, and encourage the actual effort tending to complete both rings before the end of the millenary.

A.86 and Francilienne Motorways

Fact sheet

	A.86	FRANCILIENNE
Length (total	80 km	190 km
(in service	44,5 km	132 km
Radius	8 to 13 km	20 to 30 km
Population within the circle	4,9 M. people	8 M. people

Junctions
   type
   number (total
         (in service
	grade separated
	47
		grade separated
		66
		58

Standard	Dual 2-3 or 4 lane motorway
	1-existing section
	dual 2 lane 26 km
	    – 3 – 15
	    – 4 – $\frac{3,5}{44,5}$
	2 – final situation
	dual 3 lane 45,5 km
	    – 4 – $\frac{34,5}{80\ km}$
		Dual 2 lane motorway possible widening to dual 3 lane (exceptionnally 4)

Bridges (total (number) (in service	98	198 123

Service areas
       (total
       (in service
	5 couples
	3 couples
		8 couples
		5 couples

Emergency services	1 telephone for 500 m	1 telephone for 1500 m
Traffic	see map	see map

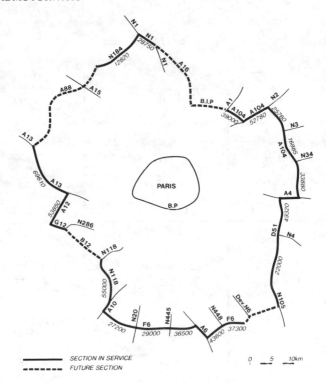

LA FRANCILIENNE (Daily traffic in 1987)

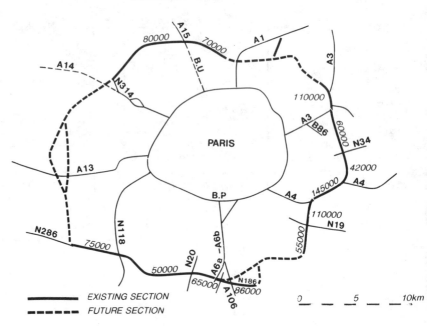

A86 HIGHWAY (Daily traffic in 1987)

The Washington Beltway

R. F. KIRBY, BSc, PhD, Metropolitan Washington Council of
Governments

SYNOPSIS. The Washington Beltway, originally designed to take
bypass traffic around Washington D.C. and its suburbs, has
become the "main street" of an ever-expanding Metropolitan
Washington Region. As part of the U.S. Interstate Highway
System, the Beltway was planned, financed, and constructed by
the federal and state governments, with 90 percent of the
funding coming from the national Highway Trust Fund. Currently, however, over two-thirds of the trips using the
Beltway are internal to the Washington region; less than
one-third are through trips. Traffic operates at stop-and-go
levels on many sections during rush hours, and major accidents are becoming all too common. Coping with continuing
strong traffic growth on the Beltway will be a major challenge for planners and policy-makers.

PLANNING AND DESIGN

1. The Washington Beltway (referred to in the Washington
Region as the Capital Beltway) was conceived in 1928 by the
National Capital Parks and Planning Commission (NCPPC), which
submitted a plan to Congress for a bypass to be built around
Washington D.C. The idea began to receive serious consideration in the early 1950s, and was finally approved by the
federal government as part of the Interstate Highway System
in 1955.

2. Construction on the Beltway began in 1956, and the first
section was opened in Maryland in 1957. The full 64 miles
were completed and opened on August 17, 1964. The Washington
Post celebrated the 25th anniversary of the Beltway on August
17, 1989 with a front page article reviewing the benefits and
problems of the Washington Region's only circumferential
freeway.

3. It is interesting to note that in June of 1964, just
prior to the opening of the Beltway, the Maryland National
Capital Park and Planning Commission (MNCPPC) proposed another circumferential freeway 162 miles long at a radius of
approximately 26 miles, compared to the eight- to ten-mile
radius for the Beltway. While this proposal received some
initial support, it was removed from regional plans during

the anti-highway era in the 1970s, only to be revived in the mid-1980s as a "Washington bypass" designed to relieve the now congested Beltway. An additional circumferential freeway, the "outer beltway," between the Beltway and the MNCPPC 126-mile circumferential was also added to regional plans at about this time. This outer beltway was partly removed and partly downgraded to an arterial during the 1970s.

4. As originally completed in 1964, the Beltway was a four-lane freeway in Virginia and a six-lane freeway in Maryland. In 1968 an impact study noted that the Beltway was stimulating extensive commercial and residential development around its interchanges, and that the majority of its traffic was already internal to the region as opposed to through travel. This study repeated the call for an "outer beltway."

5. As a result of traffic growth and new projections reflecting increasing development around Beltway interchanges, the Virginia portion of the Beltway was widened from four lanes to eight lanes beginning in 1974. In the 1980s, the Maryland portion of the Beltway has been widened from six lanes to eight lanes, with work on the last of the Maryland sections scheduled for completion by 1992 (ref. 1).

6. Only one segment of the Beltway currently remains as six-lanes with no immediate prospects for widening: the Woodrow Wilson Bridge crossing the Potomac River between Virginia and Maryland, directly south of Washington, D.C. Although widening of this bridge is currently included in the Region's long-range transportation plan, the unique ownership and operating arrangements of the bridge have somewhat delayed action on the widening project.

7. The Wilson Bridge is owned by the federal government -- the only bridge on the Interstate System that is federally-owned. The six-lane capacity of the bridge has not been increased since its opening in 1961, with the exception of the addition of break-down lanes in 1982. The bridge is approximately 5,900 feet long, with a 212-foot drawspan and a 50-foot vertical clearance in the closed position. Responsibility for maintenance and operation of the bridge is institutionally fragmented:

o   The District of Columbia maintains and operates the drawspan;

o   The state of Maryland maintains the bridge structure and roadways;

o   The state of Virginia supplies the electrical power and water service; and

o   The Coast Guard has authority over when the drawspan is raised or lowered.

8. Currently, the Federal Highway Administration, the Maryland and Virginia Departments of Transportation, and the D.C. Department of Public Works are sponsoring a design competition for widening the Woodrow Wilson Bridge (ref. 2). One of the key design issues is providing for river traffic. Currently, the bridge is opened over 400 times each year to accommodate a combination of pleasure and commercial river craft. These openings often cause substantial traffic back-ups, and as traffic on the bridge continues to grow, there are difficult trade-offs to be considered. Should the widened bridge also have a drawspan, and if so, what restrictions should be placed on the number and timing of openings? Or should the new bridge have a higher fixed span, say 70 feet, which could accommodate more river craft, but at significantly higher cost? The number of lanes for the new bridge and the possibility of including high-occupancy-vehicle lanes are also key design parameters. Various options for these parameters for the Wilson Bridge, as well as for the remainder of the Beltway, will be discussed later in this paper.

9. At the time of its completion in 1964, the cost of constructing the Beltway totaled $189 million, or $540,000 per lane mile (ref. 1). To construct the same four- to six-lane roadway in 1989 would cost $784 million, or $2.24 million per lane mile in 1989 dollars. Construction of the predominantly eight-lane freeway that currently exists would cost about $1.2 billion in 1989 dollars.

CURRENT OPERATIONS

10. The average annual daily traffic volumes on the various segments of the Beltway in 1988 ranged from 90,000 to 227,000 vehicles. The six-lane Woodrow Wilson Bridge was serving 150,000 vehicles and was frequently the site of major traffic back-ups and delays. Approximately two-thirds of this traffic is generated by the 3.4 million people living in the Washington Region (2.2 million of whom live inside the Beltway). The remaining one-third is through travel -- the travel for which the Beltway was originally designed.

11. The average speed on the outer loop (counter-clockwise direction) of the Beltway dropped from 53.6 miles per hour to 45.8 miles per hour from 1981 to 1987, due to steadily increasing traffic volumes and congestion. Speeds on the Woodrow Wilson Bridge declined more steeply than the average, falling from 50 mph to 35 mph on the inner loop, and from 47 mph to 23 mph on the outer loop. The slowest stretches of the Beltway were operating at between 10 and 25 mph during rush hours in 1987, and these speeds have undoubtedly declined even further over the past two years as traffic levels have continued to grow.

12. Accident rates on the Beltway averaged around 80 accidents per 100 million vehicle miles (MVM) in 1988, about the same rate as for all Interstate highways in Virginia, but

lower than the rate of 109/100 MVM for all Interstate highways in the urbanized portion of Northern Virginia, and lower
than the rate of 389/100 MVM for all primary highways in
Northern Virginia. While these accident rates for the Beltway compared favorably with those for similar highways in the
area, the very large volumes of traffic result in a very significant number of accidents, some of which cause major
back-ups and delays.

13. It is estimated that in 1988 the Beltway served eight
million daily vehicle miles of travel (VMT), or 16 percent of
the total VMT for the Washington Region. Over 700,000 vehicles are estimated to use the Beltway daily. Between six and
ten percent of these vehicles are trucks, and of these trucks
one-third to one-half are tractor-trailers. The accident
rates noted above translate into an average of six accidents
daily on the Beltway. Although tractor-trailers account for
only two to five percent of the vehicles using the Beltway,
they are involved in 16 percent of the accidents, or an average of one accident per day.

14. Accidents involving tractor-trailers have become a
major concern of police, fire, and traffic authorities over
the past two years. Because of their size and operating
characteristics, tractor-trailers tend to create substantial
disruptions to Beltway traffic when they are involved in
accidents. Such accidents typically result from tire or
mechanical failure on the tractor-trailer, from inattention
or error on the part of the tractor-trailer driver, or from
actions of other vehicles which may not allow adequately for
the more limited maneuverability of the tractor-trailer. On
several occasions tractor-trailers have overturned and
blocked all adjacent lanes on the Beltway for several hours.
In a few highly-publicized incidents, tractor-trailers carrying hazardous materials have overturned, resulting in fires
or dangerous emissions. These latter incidents have occa-
sionally closed down the Beltway in both directions for
several hours, resulting in massive traffic disruptions, and
requiring the temporary evacuation of nearby residential
areas.

15. In the summer of 1988, a series of serious accidents
and traffic delays on the Beltway attracted a great deal of
media attention and led to the formation of a special task
force of federal and state officials, regional transportation
and safety planners, and state and local police and fire
department officials to seek remedies for the growing problems of the Beltway. The task force appointed three work
groups to develop recommendations: one on engineering and
operations, one on education and enforcement, and one on
incident management. These work groups developed a number of
recommendations, some of which have already been implemented.
Among the latter are:

o   Installing unattended radar;

o   Requiring fuel deliveries to certain government facili-
    ties to be made at night, to keep fuel trucks off the
    Beltway in peak hours;

o   Installing additional speed advisory signs at ramps and
    interchanges;

o   Using additional portable message signs and radio advi-
    sories during major traffic disruptions; and

o   Restricting hazardous materials haulers and large trucks
    to the two right lanes.

16. Additional emergency and safety services currently
employed on the Beltway include raised pavement markings and
reflectors, the stationing of two tow trucks at both river
bridges to remove disabled vehicles, service trucks operating
constantly around the Beltway, motorcycle patrols, the use of
police cars with push bumpers, and the operation of two truck
inspection stations, one in each state. Public call boxes
were provided at several locations around the Beltway on a
trial basis, but were abandoned in favor of continuously moving service trucks that respond to vehicle break-downs or
other motorist problems.

PROPOSED OPERATIONAL IMPROVEMENTS
17. The three work groups formed in the summer of 1988 to
address the operational problems of the Capital Beltway
reviewed an extensive list of possible operational improvements. Some of the improvements identified have already been
implemented, as noted in the previous section. Of the other
improvements considered, some were adopted for implementation
over the longer term, and some were rejected as ineffective
or possibly counter-productive.
18. The engineering and operations work group recommended
the following actions for implementation over the
longer-term:

o   Introduction of an electronic surveillance and driver
    information system with incident detection, overhead mes-
    sage signs, and traveler advisory radio;

o   Review of geometric design features for specific inter-
    changes and development of plans for corrective actions
    where deficiencies are found; and

o   Review and improvements where necessary of traffic oper-
    ations along arterial and other roadways intersecting the
    Beltway.

Options that were considered and rejected included:

o Reduction of the current speed limit for all vehicles or for trucks only. This option was rejected because the current 55 mph speed limit is considered an appropriate limit for an urban freeway, and it is important that all vehicles conform to the same speed limit;

o Restriction of large trucks to the far right lane only. This option was rejected because the large number of trucks operating on the Beltway -- 12,000 daily and 600 per hour passing a single point during peak hours -- would create merging and weaving problems for all vehicles entering and leaving the Beltway; and

o A ban on all large trucks and/or haulers of hazardous materials during peak periods or at all times. This was rejected because there are relatively few alternative routes for these vehicles, and none of the alternatives that do exist are as safe for large vehicle operation as the Beltway.

19. The education and enforcement work group recommended the following longer term actions:

o Construction of additional truck inspection facilities and additional inspections of trucks and buses;

o Addition of more police patrols on the Beltway;

o A program of public information on driver courtesy and safe vehicle operation; and

o A demonstration of photo-radar equipment to detect speeders, with subsequent warning letters. (Changes in state laws would be required to permit actual ticketing by mail.)

20. The incident management work group recommended the following longer-term actions:

o Expansion of surveillance capabilities, including computerized detection from the pavement, closed-circuit television, and increased ground or aerial patrols;

o Improved communication to motorists following an incident, through the local emergency broadcast system; and

o Installation of additional water hydrants on the Beltway for use by fire services.

An option that was considered and rejected was the identification with on-vehicle placards of hazardous materials in smaller quantities than what is currently required. This was rejected because virtually all high-risk hazardous materials must already be placarded, and expanding the placarding could create confusion and decrease its effectiveness.

21. While a number of these improvement options have been considered for several years (and many are already in place on other U.S. freeways), it took the accidents in the summer of 1988 to bring an appreciation to the Washington Region of the cost of major Beltway incidents, and to create a sense of urgency about funding and implementing the improvements. A facility like the Capital Beltway is operating at such high traffic volumes, and major delays are so costly, that the relatively modest expenditures involved in these operational improvement options are easy to justify.

THE POTENTIAL FOR INCREASED CAPACITY

22. With peak traffic volumes approximating or exceeding the capacity of the Beltway, congestion and delays are currently regular occurrences. Major accidents and tie-ups are also occurring quite frequently, with traffic "horror stories" heading up the daily news at one- or two-month intervals. The Washington Region has been experiencing strong economic growth over the past eight years, and traffic volumes have been increasing at more than five percent per annum on some Beltway segments, compared to growth rates of around two percent per annum in the late 1970s. This combination of factors has stimulated increased attention to the question of increasing the capacity of the region's highway system to accommodate the growing traffic volumes.

23. The first possibility to consider is whether some of the traffic on the Beltway could be diverted to other less congested routes or controlled in a manner that reduces the chances of major delays. Unfortunately, as discussed earlier, plans developed in the 1960s for two other circumferential freeways further out from Washington than the Beltway were dropped or greatly downgraded during the 1970s. There are simply no comparable alternative routes for either the through traffic or for much of the regional traffic. Even for relatively short trips on the Beltway, the urban arterials that could provide an alternative are already congested and in need of upgrading.

24. Proposals to divert traffic currently using the Beltway or to control the access of that traffic to the Beltway are not particularly well received because of the impact they would have on intersecting or parallel roads. Road tolls and ramp metering are two possibilities that have been raised. Road tolls have received some consideration, but as a means of raising revenues for widening the Beltway rather than as a rationing device. Currently the Federal Highway Act prohibits the imposition of tolls on federally-financed roads such as the Beltway. Special legislation would be required to remove the prohibition.

25. A major effort has been undertaken to revive interest in the 162-mile circumferential freeway proposed by MNCPPC in 1964. This proposal is now being studied as two distinct Washington Bypass facilities, the portion in Maryland being referred to as the Eastern Bypass, and the portion in Virginia as the Western Bypass. The states of Maryland and Virginia have jointly funded a study of these two facilities. While there is now much more tangible evidence of demand for these roads than in 1964, there is also more development in or close to the proposed rights-of-way. This development, coupled with strong environmental concerns, places some significant hurdles and constraints on these bypass proposals. Preliminary cost estimates for these roadways are in excess of $1 billion for each facility. Presumably, funding for these roads will be sought from the federal Highway Trust Fund.

26. Over the past year or so the state of Virginia has funded a major study of the Capital Beltway in Northern Virginia, led by JHK & Associates. The results of the study are published in two volumes -- short-term and mid-term recommendations (ref. 3) and long-term recommendations (ref. 4).

27. A key conclusion of the JHK study is that even if the Washington Bypass roads are constructed, demand for travel on the Beltway will continue to grow steadily. The Bypass roads are expected to attract less than ten percent of the demand for Beltway travel, although a substantial portion of this will be for truck movement. With traffic on the Beltway growing at around five percent per annum and the Bypass facilities at least a decade away, it is clear that options must be examined for increasing the capacity of the existing Beltway facilities.

28. The JHK study concluded that daily traffic volumes on the Beltway by the year 2010 would range from 180,000 (with no Beltway widening) to well over 300,000 with improvements. Based on this demand analysis, the JHK study concluded that major expansion of the Capital Beltway was justified.

29. The short-term to mid-term recommendations of the JHK study include the following actions:

o   Pending a permanent solution to expanding the capacity of the Beltway, a major section in the southwest quadrant "should be widened to five moving lanes by reducing the width of the left shoulder, reducing the width of the travel lanes to 11 feet, and resurfacing, while maintaining the existing full-width right shoulder." It is recognized, however, that this proposal will result in some degradation in safety.

o   "A series of physical improvements involving lengthening of ramps, dualing of ramps, and implementation of auxiliary lanes;"

o   "A program of continuous lighting" to supplement the cur-
    rent lighting at selected interchanges;

o   Closed-circuit television monitoring of traffic condi-
    tions;

o   Improved signing and pavement marking; and

o   A freeway management team of highway, police, and fire
    officials to respond to major incidents.

30. Over the longer-term, the JHK study recommends major
widening and reconstruction of the Beltway to take full
advantage of the available right-of-way. The particular widening alternative recommended would involve the provision of
three separate roadways: two conventional roadways for
through traffic (three lanes) and local traffic (three lanes)
and a two-directional high-occupancy-vehicle (HOV) facility
(one lane). This amounts to a total of 14 lanes: seven in
each direction. A similar concept is being considered in
the design competition for widening the Woodrow Wilson
Bridge. The inclusion of a two-directional HOV facility is
aimed at attracting a significant portion of the
suburb-to-suburb commuter travel in the year 2010 to buses,
vanpools, and carpools.
31. A set of operational and management improvements was
recommended "to complement and achieve the full benefit" of
the 14-lane proposal:

o   A freeway surveillance and control system, including com-
    plete closed-circuit television coverage of the Beltway
    and variable message signs on the Beltway and its
    approaches to provide real-time information to motorists;

o   A complete freeway lighting and signing system;

o   Upgrading of arterial roadways that interchange with the
    Beltway, including additional lanes and more sophisti-
    cated signal control; and

o   Extensive environmental protection measures such as noise
    walls to mitigate adverse impacts on neighboring land
    areas.

32. The total cost of this extensive set of long-term
improvements is estimated to be $1 billion for the Virginia
portion of the Beltway. This is a very substantial investment requirement, and it may have to compete with other projects for federal and state dollars, such as the Eastern and
Western Bypass proposals. Further, this 14-lane proposal
would dramatically change the appearance and environs of the
Beltway, resulting in what might be described as

"wall-to-wall freeway." Whether the aesthetics or environmental implications of this proposal will be accepted by the
built-up areas around the Beltway remains to be seen.

33. A variety of other proposals have been advanced for
adding capacity in the Beltway right-of-way. The JHK study
considered other alternatives, including a 12-lane proposal
with five regular and one HOV lane in each direction. Circumferential light-or heavy- rail lines in the Beltway right-
of-way have also been proposed from time to time. And finally there are the occasional proposals for double-decking the
existing highway facility.

34. It is likely that considerable further study and public input will be required before any decision is made on
future capacity expansion for the Beltway. A host of economic, environmental, and aesthetic factors must be considered.
Clearly this decision will affect and be affected by the
types of land development that occur along the Beltway corridor. The JHK 14-lane proposal would support and encourage
continued automobile-oriented development along the corridor,
although the HOV lanes will also encourage high-occupancy
bus, carpool, and vanpool modes. By comparison, light- or
heavy-rail could be expected to encourage greater concen-
tration of development wherever stations are located.

CONCLUSION

35. The Washington Beltway has evolved since its completion in 1964 from a bypass road around Washington D.C. and
its suburbs to the new "main street" of a growing and diver-
sified Metropolitan Washington Region. Due to the continuing
automobile-oriented residential, office, and commercial
development around its major interchanges and surrounding
suburban areas, and to the lack of convenient alternative
routes for through traffic and for much of the internal traffic, the region has become heavily dependent on the safe and
efficient operation of this critical highway facility.

36. The deferment and downgrading of two other proposed circumferential freeways around Washington in the 1970s has
placed a tremendous burden on the Beltway to accommodate
growth in both internal and through traffic. Currently, the
Beltway is operating at close to or above its design capacity
during peak periods, with the result that recurrent congestion and major incidents have become serious problems.

37. A number of operational and capacity improvements have
been proposed for the short-, medium-, and long-term. Concern over congestion and incidents has provided the impetus
for action on many of the short-term, relatively inexpensive
improvements. No decisions have yet been made, however, on
more expensive medium- and long-term improvements that would
provide additional safety and capacity.

38. The Capital Beltway has had a major influence on the
development and economic vitality of the Metropolitan Washington Region. The current challenge is to ensure that it

operates safely and efficiently as the demands upon it continue to grow over the coming decades.

REFERENCES

1. DO IT COALITION. The Capital Beltway Owner's Manual. 1988.

2. BELLOMO-MCGEE, Inc. Woodrow Wilson Bridge Improvement Study. Information Brochure. November 1989.

3. JHK & ASSOCIATES, DEWBERRY & DAVIS, RHODESIDE & HARWELL. Capital Beltway Study: I-95/I-495 Northern Virginia. Short-Term and Mid-Term Recommendations Report. Prepared for the Virginia Department of Transportation. August 1989.

4. JHK & ASSOCIATES, DEWBERRY & DAVIS, RHODESIDE & HARWELL. Capital Beltway Study: I-95/I-495 Northern Virginia. Long-Term Recommendations Report. Prepared for the Virginia Department of Transportation. November 1989.

THE CAPITAL BELTWAY-WASHINGTON, D.C. REGION
APRIL 1990

FACT SHEET

Length and lanes: 64 miles, all 8 lanes except the 6 lane sections from the George Washington Parkway in Virginia to the American Legion Bridge (NW river crossing), from the American Legion Bridge to north of River Road in Maryland, and the Woodrow Wilson Bridge (south river crossing). All 6 lane sections to be widened to 8 lanes by 1992 except the Woodrow Wilson Bridge.

Radius: 8 miles North and South and 10 miles East and West.

Intersections: 38 grade separated (one every 1.6 miles on the average).

Emergency and Safety Services:
- Service trucks operate all day
- 2 tow trucks at both river bridges
- Motorcycle patrols
- Police cars with push bumpers
- Hazardous material haulers and large trucks restricted to 2 right lanes
- One truck inspection station in each of the two states
- Advisory signs for ramp speeds and interchanges
- Portable variable message signs and radio advisory (Maryland)
- Unattended radar (Virginia)
- Raised pavement markings and reflectors

Public Services: Apart from the emergency and safety services listed above no other public services or call boxes are provided.

Regional Population: 3.4 million in 1985 with 2.2 million inside the Beltway.

Cost: $189 million at completion in 1964, or $540,000 per lane mile. The same road would cost $784 million ($2.24 million per lane mile) in 1989 dollars. To construct 8 lanes for the entire 64 miles would cost about $1.2 billion in 1989 dollars.

THE WASHINGTON METROPOLITAN REGION

N

| 0 | 5 | 10 | Kilometers |
| 0 | 5 | 10 | Miles |

Frederick County

Frederick

Brunswick

Maryland

Virginia

Damascus

Montgomery County

Germantown

Gaithersburg

Leesburg

Rockville

Laurel

Loudoun County

Sterling Park

Wheaton

Greenbelt

Reston

Bowie

Dulles Airport

Washington, D.C.

Fairfax

Prince George's County

Arlington Co.

Gainesville

Centreville

Annandale

Suitland

Upper Marlboro

Fairfax County

Springfield

Manassas

Clinton

Prince William County

Dale City

Woodbridge

Waldorf

Triangle

St. Charles

La Plata

Charles County

Potomac River

Capital Beltway-Washington, D.C. Region

109

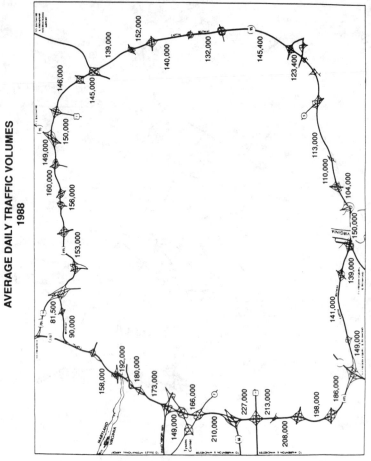

**AVERAGE DAILY TRAFFIC VOLUMES
1988**

Capital Beltway-Washington, D.C. Region

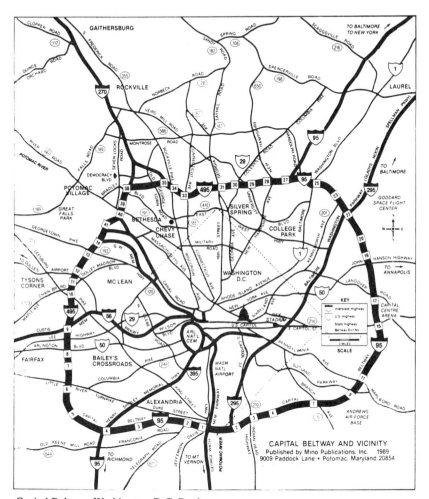

Capital Beltway-Washington, D.C. Region

Used with permission from Mino Publications, Inc.

Discussion

P.W. TEETON, Sir Alexander Gibb & Partners, Reading
I would like to refer to Transport and Road Research
Laboratory Report No. 43, 'Theoretical average
journey lengths in circular towns with various
routeing systems', which was published in 1966.
Twelve theoretical routeing systems were examined
and six basic ones are reproduced in Fig. 1. I
would also refer to the derived average theoretical
journey lengths on each of the five basic routeing
systems for three types of journey, namely:

    internal cordon journeys
    cross-cordon journeys
    and through (bypass) journeys (Table 1).

It will be seen from Table 1 that the longest
average journey lengths for internal and
cross-cordon journeys are given by the external ring
system, while that for through journeys is given by
the radial system. The shortest average journey
lengths are given, apart from the direct system, by
the external ring/radial system. The use of this
system requires, however, traffic from some
locations to pass through the centre of a circular
town to achieve the theoretical minimum journey
length. This routeing is often not feasible or
practical due to congestion, and thus the next best
practical alternatives will be sought. These are
seen to be the internal ring and/or radial/arc
systems, which are similar in concept and provide
almost equal average journey lengths.

It would appear, therefore, that to achieve the
best, or most efficient, road system for a circular
town it is preferable to have both an external and
internal ring system, with the latter positioned at
the optimum location for the mean of the journey
types. This would indicate for London, therefore,
that an internal ring at an optimum location is
required to achieve the most efficient use of the
road network.

It is instructive, to compare the radii of the

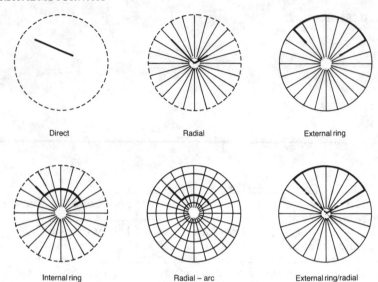

| Direct | Radial | External ring |

| Internal ring | Radial – arc | External ring/radial |

Fig. 1. Six alternative basic routeing systems for a circular town.

various orbital road systems considered at this conference (Table 2). I have no particular comment on these, except to note a marked similarity of particular radii and to ask Mr Koenig if any similar studies to those detailed in the Report mentioned above were done for Paris in studying the particular systems for that city.

M.A. SELFE, Essex County Council, Chelmsford
Mr Koenig referred to the measurement of congestion in hours/km. How is this measured in France? We are becoming increasingly concerned with providing service to the travelling public: another measure is the confidence with which a journey can be made in a particular time.

PROFESSOR P. HALL, University of California, Berkeley
I would like to ask about provision for HOV lanes (Rocades) for orbital traffic. In earlier reports these were shown as running on orbital motorways, but the latest information shows a scheme running in the southern suburbs, close to the A86, along the medians of ordinary surface streets. This may correspond to land use characteristics: the A86 runs through densely developed areas where many trips are likely to be short, while La Francilienne connects discrete new towns where trips may be longer – hence the possibility of HOV lanes on motorways. Could Mr Koenig comment on the present planning philosophy in the RIF?

114

Table 1. Average journey lengths for three types of journey on six basic routeing systems for a circular town

Routeing System	Type of journey		
	Internal*	Cross-cordon*	Through*
Direct	0.905	1.132	1.273
Radial	1.333	1.667	2.000
External ring	2.237	1.904	1.571
Internal ring	1.240**	1.534**	1.571**
Radial arc	1.104	1.381	1.571
External ring-radial	1.199	1.348	1.363

* Radius of external ring
** Based on optimum position of ring road for particular journey

Table 2. Basic orbital route statistics*

Location/orbital	Radius:km
London (pop. 8 million)	
M25	30 km
North Circular (part orbital)	13 km
Paris (pop. 10 million)	
Boulevard Peripherique	5 km
A46	13 km
La Francilienne	30 km
Washington (pop. 3.4 million)	
Beltway	16 km
Johannesburg (pop 0.8 million)	
Ring road	13 km

* From papers presented at the present conference

M.R. NEVARD, Department of Transport, London
Will 'Périphérique' reduce congestion to an
acceptable level, and what will this level be?

J.G. KOENIG
In reply to Mr Selfe, the evaluation of congestion
on the highway network is based on police reports.
This measurement is actually relying on visual
observation (by police patrols or television). In
1992, automatic data collection will be possible on
sections where magnetic detectors will be installed
under the carriageway.

In reply to Professor Hall, HOV lanes are not
actually planned on existing highways. They might
be effective in congested areas, but the control of
enforcement is very difficult without a physical
separation of this lane, as was experienced in the
case of A1 highway.

For the Francilienne highway, an HOV lane was not
yet considered, as this highway is supposed not to
be heavily congested, and as no important HOV
traffic is expected. In the case of A86, the
routing of this highway is close to the existing
RN.186 road, which is an ordinary surface street.
The relief brought by A86 to the RN.186 is being
used for implementing on the RN.186:

(a) physically protected bus lanes in the
     south-east ('Trans Val de Marne')
(b) A tramway in the north ('Saint Denis-Bobigny').

In response to Mr Nevard, the Boulevard
Périphérique reduced significantly the level of
congestion in the city of Paris when it was opened
(in the 1970s). Traffic models tend to suggest that
A86 will stabilize congestion in its influence area
during the next 5 to 10 years.

PROFESSOR R.J. BROWN, Transportation Planning Associates, Birmingham
The objective of the Washington Beltway was simply
to remove through traffic, but in fact it has had
the effect of totally altering the manner in which
the city operates. The beltway has become
Washington's main street. In the planning of such
orbital facilities we need to be far more aware of
the wider effects on urban form, and to aim more
comprehensively at achieving development as well as
transport goals.

D. BAYLISS, London Regional Transport
1.   Given that the objective of the Washington
     Beltway was to provide a bypass for long
     distance traffic, why did the planners

incorporate 38 intersections, thereby picking up almost every arterial in the corridor?

2.   The 3/4/4/3 configuration proposed works reasonably well on Highway 401 in Toronto, which has a high proportion of long distance trans-regional traffic. In Washington, with its high proportion of shorter distance traffic the weaving burden between local and express carriageways, is a 3/4/4/3 configuration operationally feasible?

P.E. NUTT, Department of Transport, Bedford
Could you comment on the organisation of maintenance operations on the Washington Beltway. Maintenance operations are a significant problem on heavily trafficked roads.

P.W. TEETON, Sir Alexander Gibb & Partners, Reading
Were traffic studies carried out for the outer bypasses, and if so, what effect will these bypasses have on existing traffic volumes on the Washington Beltway?

J. WOOTTON, Wootton Jeffreys Consultants Ltd, Woking
Has the growth in car ownership slowed down, or perhaps stopped, in the Washington DC area?

R.A. STANWAY, Stanway Edwards Associates Inc., Braamfontein, South Africa
With regard to road traffic noise on the Washington Beltway, can you give details of existing regulations, widening regulations, and attenuation measures (e.g. speed/road surface)?

R.D. RIDDETT, Travers Morgan, London
The Paper has given quite a lot of information on the origins and destination of traffic using the Washington Beltway. Can you give some details of the methods used to obtain the information, and whether, for example, vehicles on the beltway can be stopped and the drivers interviewed.?

B. PARKER, G. Brian Parkes, Twickenham
1.   Why is the purpose of the outer bypass proposals not stated as simply to relieve traffic congestion on the beltway?
2.   Where is it intended to locate the expected large increase in employment in the Washington region?
3.   Why are there difficulties in financing the bypasses, given the hypothecation of gasoline tax for highway construction?

PROFESSOR P. HALL, University of California, Berkeley

Can you comment on the role of development agreements in financing highway improvements in the outer part of the Washington DC area?

DR G.T. LATHROP, Department of Transportation, Charlotte, USA

In response to a comment regarding the disposition of 'all the money from user fees' in the USA, I would emphasize that the fines certainly are not adequate to the needs for additional or improved transport infrastructure.

In response to a comment about privatization, I would indicate that this is a complex process in the USA, particularly when the point is to obtain funding participation in joint public-private projects. Complications include: multiple, diverse circumstances in the 48 states; whether projects are on- or off-site; when the project is to be built, etc. Developers resist participation: one termed public attempts to get funds as exaction, extraction, extortion.

G.J. HILL, Travers Morgan One Ltd, East Grinstead

UK motorways operate at 70 mile/h speed limit (which is not effectively enforced) and with land discipline requiring overtaking on the right hand side only. (This is tending to break down as congestion increases and results in uneven lane utilisation). To what degree does the Washington Beltway benefit in capacity and safety terms from a lower enforced speed limit and 'pass on either side' lane discipline?

M.R. NEVARD, Department of Transport, London

Tractor traders constitute a major economic cost, comprising 2-5% of traffic, 16% of accidents, and probably a higher percentage of accident induced delays. What measures are you taking to deal with this in relation to drivers, vehicles and size?

S.N. MUSTOW, Consultant, Sutton Coldfield

You said that you wished you could see a comprehensive land use and transportation plan for Washington DC under which beltways could be prescribed. How can this be achieved?

J.M.H. KELLY, Rendel Palmer & Tritton Ltd, London

I noted from the Washington Beltway presentation that the consultants looking into ways of increasing the capacity of the Washington Beltway proposed narrowing lanes and reducing hard shoulder widths so as to make more use of the available road space. This was also one of the proposals put forward for

short term improvement of the M25, but met with
general opposition.

I would like to know the reaction to such ideas in
Washington, and if there has been debate in Paris on
whether carriageway lane widths and hard shoulder
widths can be sensibly reduced to maximise use of
road space.

J. WILLIAMS (On behalf of R.F. KIRBY)

In response to Mr Bayliss, the original plans for
the Capital Beltway assumed a limited number of
interchanges, since the Beltway was thought to be
primarily for through travel. Getting the Beltway
built required negotiations and compromises with
state and local governments, and the various
interests which they represent. In the course of
these negotiations, the number of interchanges
multiplied, as planners and politicians sought to
use the Beltway to address transportation needs
other than serving through trips. The decision to
build further interchanges was made easier by the
federal government, which paid 90% of the cost of
the Beltway.

Regarding the second question posed by Mr Bayliss,
the 3-4-4-3 concept for the Beltway has been
proposed by consultants, JHK and Associates of
Alexandria, Virginia. In their words:

'This concept has seven continuous lanes in each
direction, four continuous lanes in the inner
roadways (express lanes) and three continuous lanes
in the outer roadways (collector lanes). The
majority of merging, diverging and weaving
associated with interchange ramps would occur on
the outer roadways. An HOV lane in each direction
could operate in the innermost lane of the inner
roadway. A minimum four foot buffer would be
provided between the HOV lane and regular lanes.
Separate ramps would provide direct connections to
other HOV facilities in the region.
Slip ramps are provided in many locations, often
grade separated (commonly referred to as braided
ramps) to allow the movement of traffic from the
inner roadways to outer roadways and vice versa.
Slip ramps are integral components of a
collector-express freeway system. Generally with
this type of design, the express roads are intended
to serve the longer distance trips while the
collector roads serve local traffic and connect
directly with all interchanges.'

In response to Mr Nutt: in Northern Virginia,
maintenance work for the Beltway is divided into two

categories: 'ordinary' (filling pot-holes, cleaning trash, mowing grass, etc.) and 'maintenance/replacement (resurfacing, replacing joints etc.). Ordinary work is done by state employees during the 10 a.m.-3 p.m. period. Maintenance/replacement work is done by contractors from 9 p.m.-5 a.m. This timing is chosen to cause as little disruption as possible to traffic flow. One exception is snow removal, which is done regardless of time of day.

In response to Mr Teeton, traffic studies have been conducted. They show that the outer bypasses will provide some relief for congestion on the Beltway, particularly through diversion of truck travel. Neither the Eastern nor the Western Bypass will solve Beltway congestion, and Beltway traffic volumes are forecast to increase, even if the bypasses are built.

In reply to Mr Wootton, the rate of auto ownership for metropolitan Washington is currently 1.8 vehicles per household. This is expected to increase to 2.0 vehicles by the year 2010. The actual rate of growth will decrease somewhat in this period.

In reply to Mr Stanway, currently, freeways (motorways) are candidates for federally funded noise abatement measures whenever capacity is added through widening or new construction. In these cases, the Federal Highway Administration uses a highway noise prediction model to forecast impacts. In the case where residences are close to the road in question, if forecast highway noise plus ambient noise is greater than Leq (1 hour) 67 dBA outside the house, or Leq (1 hour) 52 dBA inside the house, noise abatement measures must be employed. These measures include acquisition of property for buffer zones, installation of noise barriers, and sound-proofing existing structures.

In response to Mr Parker, as with any transportation facilities of this magnitude, the bypasses have a number of purposes. One stated purpose of the outer bypass proposals is indeed to relieve congestion on the Beltway. As noted above, traffic forecasts have shown that the bypasses will be only partially effective to that end.

Growth in employment is expected to occur largely in the downtown core, around Metrorail stations in radial corridors and in new suburban centers.

The Eastern and Western Bypasses should each cost approximately $1.5 billion. This, added to the cost of the Washington region's other 20-year transportation construction needs, far exceeds anticipated revenues dedicated to that purpose, including the gasoline tax.

In response to Professor Hall, impact fees, proffers and tax districts are all used to secure private sector financing of highway improvements in the outer areas. The total revenues from these sources are still a small proportion of our total bill for transportation improvements. One limitation of developer agreements is that they can only reasonably be applied to the vicinity of the development in question, whereas the travel impacts of the development may be widespread.

In reply to Mr Nevard, as noted in the paper, truck accidents are a major problem for Beltway operations. A Freeway Management Team has been formed to manage major incidents such as truck accidents. The team includes state and local transportation officials, state and local police, and fire departments. Actions pertaining specifically to trucks include restriction of trucks carrying hazardous materials to the outside lanes (the left lanes in the UK); the construction of new truck inspection stations; and scheduling of fuel deliveries to the evening hours, to keep fuel trucks off the Beltway in peak hours.

In response to Mr Mustow, one way to achieve a comprehensive metropolitan land use and transportation plan would be to gather all appropriate government bodies, environmental interests and business interests into a single, decision-making group to develop a vision of the future and a means to implement the vision. In Washington, we have begun this process with the creation, in June 1990, of an advisory group, the Joint Task Force on Growth and Transportation.

In response to Mr Riddett, origins and destinations for select links of the Beltway and other major highways are obtainable from the regional travel forecasting process at the Metropolitan Washington Council of Governments. This computerized process simulates travel on the highway and transit networks. The process is periodically calibrated using original-destination information from home interview surveys and the US Census. Special surveys of trucks or through travel are sometimes conducted using licence plate information gathered by roadside surveillance. It is generally not accepted practice to stop traffic for interviews on freeways (motorways).

In response to Mr Hill, the 'pass on the right only' rule might result in reduced safety and highway capacity, due to the large amount of lane-changing which it produces. To fully resolve this issue would require a detailed traffic engineering study.

In reply to Mr Kelly, we have no experience in Washington with reducing freeway lane widths and shoulders, except on the radial Shirley Highway outside the Beltway, where one shoulder is being used as a temporary HOV lane during the peak period. There is currently no consensus on whether or how these practices should be employed to add capacity. There may be major negative impacts on enforcement and safety; removing disabled vehicles would certainly be more difficult. There is a project to address this issue which will be performed by the National Cooperative Highway Research Program of the Transportation Research Board, scheduled to start this winter.

The Johannesburg National Ring Road

M. F. MITCHELL, PrEng, BSc (Eng), MAdmin, FIHT, FSAICE MITSA, MCIT, Department of Transport, L. M. G. P. LUCYKX, PrEng, BSc (Eng), MSAICE, MITE, MITSA, MSAConsE, Scott and de Waal Inc., and R. A. STANWAY, PrEng, BSc (Eng), MSc (Eng), DIC, FCIT, FITSA, FSAICE, MICE, MSAConsE, MSAIMunE, Stanway Edwards Associates Inc.

SYNOPSIS
    This paper traces the history of the Johannesburg National Ring Road from its conception around 1945 to its full opening in 1986. Aspects such as design, construction, maintenance and transportation and land use development are covered followed by details of current trip characterstics, traffic flows, operations and monitoring. Road traffic noise and alleviation techniques are also discussed and overall conclusions are drawn about possible short-term operational and long-term infrastructural features for consideration.

INTRODUCTION
    1. Despite a great deal of scepticism amongst certain persons, engineers of the South African central government roads authority started a process of reservation of land for a corridor for the Johannesburg Ring Road in the late forties, in an area which at that time was farming land surrounding, and removed from, the then boundary of Johannesburg.
    2. This far sighted action bore fruit some 20 years later with the start of the construction of this very valuable road facility, which has played a major role in traffic distribution and land usage in the present Johannesburg metropolitan area.
    3. The Ring Road has been open to traffic in its entirety for only four years, though certain sections have been under traffic for 18 years, and it is reaching capacity over certain sections and at specific interchanges.
    4. Attention is being given to the application of traffic management principles and isolated capacity improvement techniques to put off the day when major works will be necessary to increase its ability to cater for the traffic demand.

BACKGROUND
    5. Johannesburg, which lies at the centre of the Witwatersrand, the financial and industrial epicentre of South Africa, is the economic capital of the country. The

central Witwatersrand or Johannesburg Metropolitan (JOMET) Area, covers an area of approximately 50km by 50km with a population of some 3,5 million (see Fig. 1).

6. Johannesburg itself has a population of some 2,0 million persons and incorporates a central area with many tall and closely spaced office blocks and flats. It unfortunately has a land use pattern which is not conducive to efficient road transport. Travel patterns in Johannesburg are extremely dispersed with generally speaking large travel distances between home and work places. The private motor car plays a major and ever increasing role in these travel movements despite much attention having been devoted during recent years to encouraging the use of conventional public transport, mainly the bus mode. The road corridors in the metropolitan area have traditionally been unbalanced, with the north-south direction being far better served than the east-west direction.

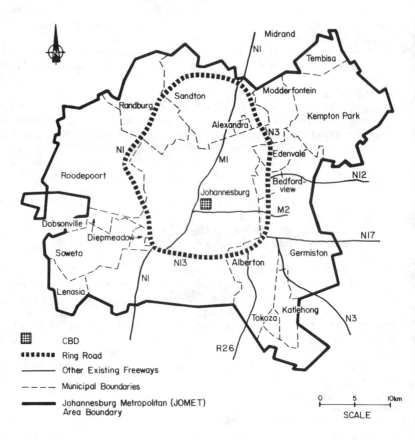

Fig. 1. Johannesburg metropolitan area and the National Ring Road

7. Administratively the responsibility for the provision of roads within this area resides in three levels of government. Firstly the South African Roads Board, as the central government road authority, has the responsibility for the provision of national roads. These roads are intended to form the major road traffic links between important cities and the various regions and growth points in the country. The regional road network is provided by the provincial authorities with local authorities, often grouped into metropolitan transport authorities being responsible for city and metropolitan roads, including intra-metropolitan arterials.

THE RING ROAD CONCEPT

8. The Johannesburg Ring Road was constructed as a national road and comprises three separate routes, viz. the N1, the N3 and the N13 as illustrated in Fig. 2 together with its interchanges and associated existing and planned future freeways. Among the road-building issues which have caused heated controversy over many years was the question of whether or not national roads should bypass towns. The controversy started during the forties and continued well into the sixties, when the building of freeways in metropolitan areas particularly, became an important issue. Engineers of the predecessor of the South African Roads Board, as far back as 1938 firmly believed that national roads should be built around, and not through towns and villages if they were to serve their purpose as routes for long distance, fast-moving traffic in the years to come. On the other hand many municipal engineers and town councils strenuously opposed this concept arguing that travellers would wish to pass through the towns concerned and also that such a move would disastrously affect commerce in the smaller towns.

9. Much political manoeuvering took place during the forties and fifties, with engineers of the central government road authority arguing that even if the time was not ripe to build these bypasses at that stage, the least that should be done was to acquire the necessary road reserve to do so at a later stage.

10. The notion of national ring roads around the major cities was an extension of the bypass concept. The Johannesburg Ring Road proposal started in 1945 when the South African Road Board´s planning engineer foresaw the rapid expansion of urban development around Johannesburg and realised that any prospect of building the Ring Road in the future would be destroyed unless a right-of-way could be reserved.

EARLY PLANNING

11. The first reservation of a section of the eventual Johannesburg Ring Road was done during the late forties with subdivisions of land at Bedfordview on an informal basis by

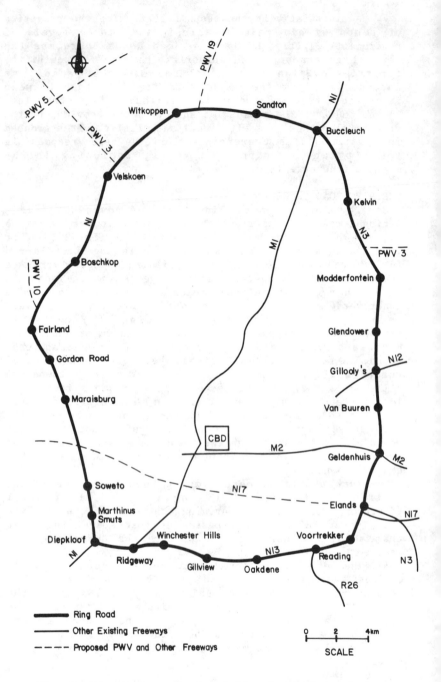

Fig. 2. Johannesburg National Ring Road, interchanges and associated freeways

the Road Board's planning engineer, with the assistance of the Transvaal Surveyor General, an old university friend, by the reservation of a corridor 180 Cape feet (56,7m) wide. He verbally requested the Surveyor General to reserve the 180 Cape feet corridor according to rough sketch plans for a possible future national road.

12. Other sections to the north of Johannesburg were later reserved according to rough plans showing only the road centreline, during the same period. In certain instances part of the conditions for the approval of township development was the reservation of a 180 Cape feet reserve for a future national road even though the then existing national road passed through the centre of Johannesburg from north to south.

13. During the same period surveys were carried out, and where township development was considered likely, sections of national road were proclaimed after clearing areas of mine reservations with the Department of Mines. These road proclamations were done on a road centreline basis only. No design plans were available and no reserve widths were shown or proclaimed.

14. These actions however established the principle of making provision for the future Ring Road and this bore considerable fruit in the future as it had been possible, for example, to avoid areas of shallow undermining for gold, and very costly property for expropriation. Many other important benefits relating to industrial, gold mining and commercial development accrued as they were able to be optimally located from the viewpoint of the Ring Road.

15. Despite this early foresight, so much opposition was encountered to the planning of a national Ring Road to avoid the then metropolitan Johannesburg, that progress eventually came to a standstill. In 1958 however the scheme was reviewed and consulting engineers were appointed to undertake the detailed design.

DESIGN, CONSTRUCTION AND MAINTENANCE

16. Consulting Engineers were appointed for all phases of the design and the compilation of contract documents. Construction was carried out entirely by private sector contracting firms, with the work being awarded on the basis of open tenders. Site supervision was generally done by the consulting engineers appointed for the design. There was thus a high level of private sector involvement, with the client exercising overall control.

17. The first contract on the Ring Road was for a 12km section on the eastern side, between the Buccleuch and Gillooly's interchanges. This contract was awarded in 1966 and comprised a dual carriageway, four-lane freeway at a cost, at that time, of 3,75 million Rand (1,9 million Pounds at the exchange rand then prevailing). The work on the full eastern section (N3) was completed in 1978 providing 12km of four-lane freeway and 13km of six-lane freeway.

18. Design of the southern section of the Ring Road (N13) known as the southern bypass, started in 1961 and the first contract for work on this project was awarded in 1976. The entire southern bypass was opened to traffic in 1986. Design of the western section of the Ring Road, the N1 section, began in 1965 with the construction taking place between 1975 and 1982.

19. It was estimated in 1986, when the Ring Road was completed, that the total cost at prices ruling when the work was carried out was 250 million Rand. The replacement value of the Ring Road at current prices exceeds twice that figure at least.

20. The design speed of the Ring Road is 60 miles per hour. The topography is gently rolling, though in some areas hard quartzite ridges had to be traversed. A great variation of geological conditions prevailed with some of the more significant obstacles including shallow undermining (gold), construction through old mine dumps comprising fine silt (slimes dams), and marshy areas.

21. The pavement comprises basically two types, namely; a jointed, reinforced concrete pavement, without dowels (except where poor road subgrade conditions prevail), over a lean mix concrete subbase, and, hot rolled asphalt wearing courses of up to 150mm thickness over cement stabilised crushed quartzite bases and stiff stabilised subbases.

22. Dynamic substitution. This technique was necessary to improve the subgrade conditions through marshy areas on the western section of the Ring Road. The materials on which the roadway had to be placed consisted of 6m of marsh sediments and mine slimes overlying the bedrock, and were not capable of carrying the 9m of fill required for the road. After discarding more conventional options, such as excavation and replacement of the unsuitable material, and a viaduct on slim piles at close centres, it was decided to opt for the dynamic substitution method.

23. This method entailed the driving of rock columns through the slimes and marsh onto the bedrock. Initially a rock blanket was tipped to a depth of approximately 1,5m over the whole area. Rock columns of 3,5m diameter at 7m centres were created using a 45m high Menard tripod dropping a weight of 38 tons through a height of 27m. The depressions in the rock blanket caused by the falling weight were refilled with rock and repounded, until the columns were seated on the bedrock.

24. Three level diamond interchange. One important interchange on the Johannesburg Ring Road is of particular interest. This is known as Gillview (named from a nearby suburb) and it is centrally situated on the southern section of the Ring Road on a cross road called Kliprivier Drive.

25. The interchange is constructed in the form of a three-level diamond with interchanging traffic using the middle level (see Fig.3). This middle level has the north-south roads in addition to the terminals of the Ring Road

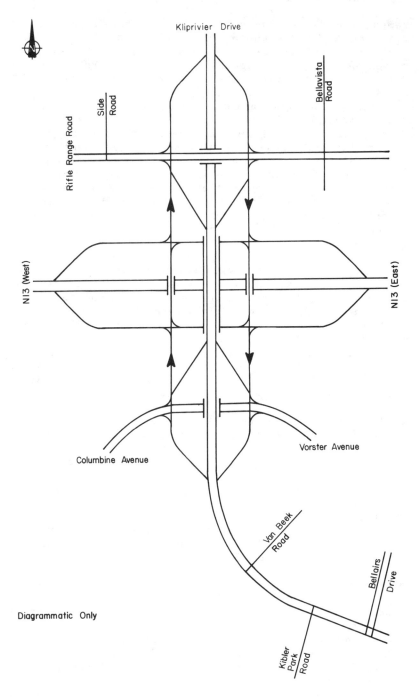

Fig. 3. Layout of the Gillview Interchange

ramps. The result is nearly two kilometres of collector-distributor road running parallel to a central dual carriageway limited access road. At present the interchange serves as an access interchange to the Ring Road; in the future a freeway route may connect to the central dual carriageway.

26. The three level diamond configuration was chosen for four reasons. Firstly the topography of the area enabled the road gradient on the two important through routes to fit comfortably without major earthworks in rugged terrain with solid rock outcrops; secondly the existing important side roads could be linked into the corridor; thirdly the interchange layout is compact and hence property acquisition was reduced; and fourthly the high traffic volumes on Kliprivier Drive (4 800 vehicles/hour in one direction) through the interchange required special attention to be given to adequate lane provision.

27. Maintenance of the Ring Road by private contractors has been to a high level with regard to pavement, roadside furniture and grass cutting within the road reserve. The pavement layers have been maintained and strengthened during upgrading to six lanes on certain sections of the road.

28. Maintenance of the signal systems on the access interchanges is by agreement with the local authority for the first level of maintenance or by a maintenance agreement directly with the signal manufacturers. The road signs on the Ring Road are at present being maintained and replaced under a separate road signs contract.

DEVELOPMENT OF THE METROPOLITAN AREA

29. Land use. The Ring Road has been one of the most important catalysts influencing the decentralization process in the metropolitan area. Residential, commercial and industrial land uses have been attracted by the increased accessibility provided by the Ring Road close to the access interchanges. These different land use types have symbiotically influenced the development of one another, with residential development usually being the forerunner.

30. Although the satellite towns of Randburg and Sandton have always been viewed as areas for expansion, the rate at which this took place has been accelerated by the presence of the Ring Road. Commercial development in the north-west of the area in particular is at an advanced stage. To the south of the Ring Road the residential boom is already in full swing and commercial development can be expected to follow shortly. To the east of the Ring Road, where a fair degree of residential development had already occurred, property values have risen sharply since the opening of the road to traffic.

31. The Ring Road has also provided increased accessibility to the major north-south transportation corridor between Johannesburg and Pretoria, thereby contributing greatly to the strengthening of this

development axis. There has been rapid commercial and residential development in this corridor as evidenced by the rapid growth of the town of Midrand, which only became a municipality in its own right recently.

32. The development which the Ring Road has initiated, particularly in the north-west of the metropolitan area, is not without its drawbacks however. The development which has taken place is very much oriented to the use of private transport.

33. Individual high income residential plots are large, generally of the order of 1300 square metres. There is very little high density residential development in these areas, and the little that there is occurs in isolated small pockets rather than in concentrations along arterials which could be served by public transport. Metropolitan planners, who have as one of their major objectives the promotion of public transport among the high income residents, find their task in this respect made more difficult due to the accessibility provided by the Ring Road.

34. Low income residential areas with good accessibility to the Ring Road are generally of high density but they have predominantly low car ownership at present with public transport being the main mode.

35. Transportation. Since the 1950´s planners in Johannesburg have been predicting congestion of monumental proportions on the city´s roads unless long-term solutions to the problem of growing traffic volumes were implemented. This concern is reflected in the fact that by 1955 Johannesburg had plans for an extensive network of freeways and high standard arterials, and, inter alia, in 1971 commissioned a feasibility study for an underground rail based mass transit system.

36. Implementation of these plans was hampered by the city´s inability to raise the necessary funds and public opposition. By the early 1970´s, Johannesburg had completed, with some assistance from central government, only the core of its freeway system in the form of two freeways serving the central business district (CBD). The M1 runs in a north-south direction and the M2 east-west (see Fig.2). Both of these freeways provide access to the CBD and allow traffic to bypass the CBD. Although the M1 is continuous within the confines of the Ring Road, the M2 terminates at its western end before reaching the Ring Road.

37. Subsequent to the construction of the M1 and M2, no major addition to the transport infrastructure of the city has been made. Implementation funds were not available, and central government, despite numerous pleas by Johannesburg, was not in a position to provide assistance.

38. According to the predictions, Johannesburg should have stared disaster in the face in the light of all the facilities which had been planned but not implemented, such as metropolitan freeways and arterials and a modern public transport system. The construction of the Ring Road together

with major land use decentralization played a major role in averting the predicted congestion.

39. The Ring Road's role was twofold in this regard. Firstly, it to a degree provided alternative routes to the city's freeways and arterials, and thus reduced the pressure on them. While these ring road routes may have been longer in distance, users' assessment of the generalized cost of travel, particularly in terms of time and convenience, ensured that they would be used.

40. Secondly, the Ring Road promoted the decentralization of activities away from the CBD by the provision of accessibility to previously undeveloped areas. The predicted expansion of activities and consequent increase in total travel did occur, but a major proportion of this was accommodated in the area opened up for expansion by the Ring Road, and did not require to be accommodated on the city's limited facilities.

41. Transportation-related land use development in the whole Witwatersrand has been influenced by the Ring Road and its associated freeways to the extent that private transport commuters can live virtually anywhere in the region and still be reasonably accessible to their workplaces.

TRAVEL PATTERNS AND OPERATIONS

42. **Trip characteristics.** The predominant trip purposes of traffic utilizing the Ring Road on weekdays are commuter trips (home-based work) in the peak periods, and work-related, including freight trips, in the off-peak periods. A significant contributory factor in this regard is the accessibility of the Ring Road to the majority of the major employment centres in the region, normally via other freeways linking into, and across, the Ring Road. In addition, as the majority of the arterial roads are radially orientated, with a limited supporting road network parallel to the freeway, a number of relatively short circumferential trips occur.

43. The unique network of freeways connecting to the Ring Road has emphasized this trip pattern where another of the roles of the Ring Road is clearly at present to distribute traffic circumferentially to the next closest interchange and so link up with another freeway or major arterial radially connected to the Johannesburg CBD.

44. As the major portion of the traffic using the Ring Road on weekdays is primarily metropolitan-generated, estimated as being between 90 per cent (off-peak) and 95 per cent (peak period) of the total traffic, and due to the occurrence of peak period congestion on the facility, extensive discussion has taken place amongst members of the Southern African Roads Board as to whether some restrictions, for example, ramp metering, should be placed on the use of metropolitan traffic on the Ring Road, in order to avoid the prime use of national routes i.e. to cater for long distance traffic, from being subverted.

45. The other major trip purpose of traffic on the Ring Road is leisure-related trips, occurring mainly on Friday and Sunday evenings, as well as on long weekends and at the beginning and end of school holiday periods. In this regard the Ring Road links into the national route network, providing accessibility between the metropolitan areas in the vicinity of Johannesburg and the major holiday resorts.

46. Traffic flows. The seven day traffic volumes in 1987 on the Ring Road are indicated in Fig. 4. The range of these volumes is between 23 000 and 41 000 on the two-lane carriageways, and up to 66 000 on the three-lane carriageways (two-directional). The proportion of heavy vehicles

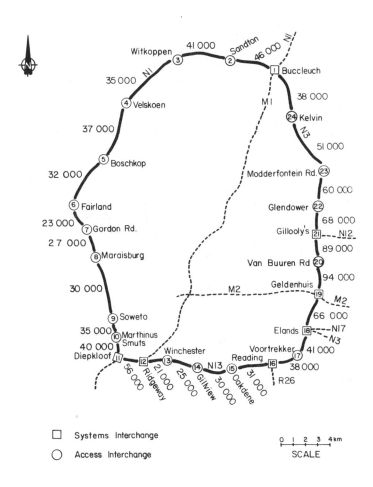

Fig. 4. Seven day average traffic flows on the Ring Road in 1987

ranges between five per cent and 20 per cent. The high level
of loading on the Ring Road is indicated by peak period flow
rates of 2 200 vehicles per hour per lane, with complete
flow breakdowns occurring on certain sections. Morning and
evening peak hour factors range betwee 0,93 and 0,98.

47. A notable feature of the traffic flows are the high
volumes on the access interchanges, with consequent
congested conditions occurring at the ramp terminals and at
the ramp/freeway junctions. This is due to the extensive use
of the Ring Road for commuter trips, and the large catchment
areas of the interchanges, due in turn to the significant
spacings between interchanges (ranging between 1,2km and
5,8km). On some of the three-lane carriageways, for example,
the ramp flows represent one-third of the total flow
downstream of the ramp/freeway junction. The maximum
recorded hourly flows are 1750 on the on- and off-ramps.

48. High volumes of cross-freeway traffic also occur at
the interchanges, conflicting with the abovementioned ramp
volumes and resulting in significant congestion, with
potential queueing conditions back onto the freeway proper.

49. An interesting feature of the traffic flows at the
ramp terminals is the occurrence of different peaking times
for conflicting movements. This arises when there are major
residential and employment generators in the immediate
catchment areas of the interchange, resulting in the morning
period on-ramp movement peaking earlier (between 07:00 and
07:30) than the off-ramp movements (between 07:45 and
08:00). These times are related to the distance of the
interchange from the final destination i.e. the work
location. This situation lends itself to multi-plan
signalized operations in the peak periods, in order that
such interchanges can be efficiently operated.

50. The annual traffic growth rates on the freeway have
been high over the last five years, i.e. up to six per cent
on certain freeway sections. It is however difficult at this
stage to assess longer-term trends on the whole freeway, as
the southern portion of the Ring Road has only recently been
completed.

51. Operations. Enforcement on the freeways is undertaken
by the traffic departments of the eight local authorities
through which the Ring Road is routed. This includes both
enforcement at the ramp terminals as well as on the freeway
sections.

52. With respect to incident servicing, SOS emergency
telephones have been installed on the Ring Road, located at
approximately 2km intervals. The system is operated by the
Automobile Association which in turn has communication links
with ambulance services, traffic departments, police
stations and breakdown repair services.

53. With respect to intersection control at ramp
terminals, detailed cost-benefit studies have been
undertaken, with the assistance of simulation programs, on
the relative cost-efficiency of the various control

procedures i.e. priority, pointsman and signal control. Irrespective of the economic viability, with the danger of queues extending back from the ramp terminals onto the freeway proper, the installation of traffic signals is more often than not justified.

54. Furthermore, due to the unique traffic nature of interchange operations, local signal warrants have been developed. These are based on the number of 15-minute time periods that non-priority movements are delayed by more than 30 seconds on average. This local warrant requires four such congested periods per day to justify signal installation.

55. A number of traffic control measures and strategies have been adopted at the interchanges, namely;

(a) multi-plan operations within a peak period, to cater for the aforementioned different times of peaks for conflicting movements,

(b) double right-turn lanes operating under gap acceptance conditions. Although this is not accepted by certain local authorities, accident statistics have indicated that such operations have not resulted in increases in accidents,

(c) dropping of exclusive turning stages during the off-peak periods, and

(d) reduction of lane widths to 2,9m on the ramps and the cross road, in order to introduce additional lanes in conjunction with use of the road shoulders.

56. The above measures have resulted in the ramp terminal capacities being optimized, and the balancing of capacity requirements between the ramp terminals, ramp/freeway junctions and the freeway proper are presently receiving attention.

57. Monitoring. There are three monitoring activities undertaken relating to the operations on the Ring Road, namely;

(a) the comprehensive traffic observation (CTO) programme

(b) the monitoring of the Ring Road traffic, and

(c) the monitoring of interchange operations.

58. The CTO programme, which is undertaken on the national freeway network, has 11 stations on the Ring Road. Each station is counted for a 48-hour period each year, and volumes, vehicle classification, speeds and headways are recorded. The objectives of this programme are to derive management and design information, and monitor trends on the national freeway network.

59. The Ring Road monitoring programme, which is undertaken by the University of the Witwatersrand, includes seven-day counts on each freeway link and ramp, as well as a monthly seven-day count at five control stations on the Ring Road. This programme provides useful information on the Ring Road operations and has as its overall objective, the monitoring of trends on the freeway.

60. The interchange monitoring programme was initiated in 1982. This was in response to the increasing frequency of

complaints concerning congestion received from the driving public and local authorities. The objective of this monitoring programme is to undertake short-term planning on a pro-active basis, in order that potential problem areas can be addressed before they are manifested at the interchanges.

61. The general planning procedure includes the following steps, namely;

    (a) the undertaking of peak period turning movement counts and the assessment of the existing volume-capacity ratios,

    (b) a course assessment of growth rates at interchanges, based on either catchment areas growth rates and/or historic growth rates,

    (c) the application of growth rates to the existing volume-capacity ratios, from which an indication of the spare life of the facility can be obtained,

    (d) based on this residual capacity, a future monitoring programme is derived with the primary objective of identifying pending problem areas approximately three years before modifications are required, and finally

    (e) these growth rates are then monitored to detect any major deviations from those applied in the analyses.

62. This programme has proved to be successful in terms of pre-empting complaints from the public, or else providing a quick response to implementing upgrading measures.

ROAD TRAFFIC NOISE

63. Numerous complaints regarding the noise generated by the Ring Road have been received over the years since its construction, but by the mid-1980's the situation was found to be worsening with the rapidly increasing traffic volumes and specifically the increased heavy vehicle usage of the facility. Although the whole Ring Road was found to have a noise problem, the worst impact is experienced along the concrete surface sections in the north, west and south, that is, on N1 between the Diepkloof and Buccleuch Interchanges (Johannesburg Western Bypass) and N13 between the Diepkloof and Reading Interchanges (Johannesburg Southern Bypass).

64. Because of the recent promulgation of new environmental legislation, a detailed investigation was initiated to identify the worst future noise situation along the concrete surface sections and then to assess the cost implications of instituting noise attenuation measures.

65. Using the South African Bureau of Standards prediction model it was found that for the future worst situation, over 95 per cent of the 38km length of the Western Bypass exceeded a 65dB(A) noise level at the road reserve boundary. Along the 10km length of the Southern Bypass the 65dB(A) level was exceeded at all but two of the locations modelled. The noise level at the road reserve varied between 64dB(A) and 72dB(A).

66. The next stage of the investigation evaluated and costed several practical means of attenuating the noise along a representative 10km section of the Western Bypass, namely;
(a) reduction in the average operating speed along the freeway,
(b) resurfacing of the concrete with a rubber bitumen asphalt overlay, and
(c) the construction of noise attenuation barriers.

67. The reduction of the average running speed from 110km/h to 90km/h was predicted to reduce the noise level by only approximately 1,5dB(A). It would be effective in reducing the noise level to 65dB(A) at the reserve boundary in very few areas and could therefore not in itself be considered as a long-term solution to the problem.

68. Tests on gap graded asphalt overlays of concrete freeway have indicated that a constant reduction of 5dB(A) can be achieved by this method. The long-term economic and acoustic viability of this overlay method is however still being researched. This procedure would not reduce the noise level to 65dB(A) in all the areas along the study section of the freeway and would have to be supplemented by about 2,6km of barrier. (This is approximately 17 per cent of the total barrier length which would be required if the construction of barriers were to be the only noise attenuation method applied). Attention however is to be given to open graded asphalt overlays as a noise reduction measure and it is hoped that this will lessen the problem.

69. Along the 10km section of freeway, approximately 15,3km of barrier (inclusive of both sides of the freeway) would have to be constructed to meet the 65dB(A) requirement at the road reserve boundary.

70. Seven barrier systems were investigated and costed. The use of earth berms was also considered but not evaluated in any detail as the acquisition of additional property in the residential areas which flank most of the study section of freeway was not considered economically or politically viable.

THE FUTURE

71. The future development of the Ring Road will depend to a very large extent upon the allocation of short- and long-term responsibilities between the South African Roads Board and the metropolitan and local authorities through which the facility passes. This is because the Ring Road is clearly currently serving a predominantly local function but it is being funded, operated and directed from a national level.

72. Independent of major upgrading options, there will be a need to maintain and operate the Ring Road optimally now that it is complete and travel patterns are stabilizing. This should be based upon a monitoring programme of pavement/structural aspects on the one hand and operational aspects on the other hand.

73. As far as operational aspects are concerned, the basic criteria are safety and level of service and the monitoring will be able to identify appropriate steps to be taken. In the case of safety, measures such as striping, signage, lighting, patrol cars, road surfaces etc., will in all probability require attention while in the case of level of service, measures such as freeway traffic management systems have already started to be considered.

74. Freeway traffic management systems in general have been examined for a very short section of the Ring Road on both sides of the Buccleuch interchange and these have been reduced to freeway traffic control systems such as the possible exclusive reservation of shoulder lanes for on- and off-ramp traffic together with variable message signing. Ramp metering has been excluded at present as ramp volumes exceed those that are possible to control in practice. This is due mainly to the large distance between interchanges i.e. national road design standards for a freeway now operating as an urban freeway.

75. In the longer-term, consideration may have to be given to a metropolitan-wide traffic management system linked to the Ring Road operation so that the ramp metering concept can be fed back along the many arterial roads on an area-wide basis. All of the above measures however depend upon the reaction and behaviour of Southern African motorists under freeway traffic management conditions and an experimental pilot project is planned to go into operation shortly to assess acceptance.

76. In the long-term, major and possibly highly controversial measures will need to be investigated and these may include the likes of road pricing through tolls; major road widening; the introduction of intermediate interchanges; service roads; major interchange upgrading and finally, even new national route locations for the N1 and N3.

77. Associated long-term possibilities are the construction of the planned freeways indicated in Fig.2 which will alter travel patterns and further influence land use changes, particularly those that extend to within the Ring Road.

78. In the overall context, both environmental and land use aspects have already come under the spotlight and pilot implementation of road traffic noise attenuation measures is likely to take place. The land use aspect is also being addressed via a cost-apportionment study as it relates to land use development immediately adjacent to national road interchanges while the broader aspects of travel patterns and land use are the subject of an ongoing metropolitan land use/transportation study.

JOHANNESBURG NATIONAL RING ROAD

Fact Sheet

GENERAL INFORMATION

Length
79,6 km (49,5 miles)
Total length of concrete $\pm$ 50km
Total length of asphalt $\pm$ 29km

Radius
Generally 28 kms (17 miles)
25 kms between Glendower and Gordon Road interchanges
32 kms between Sandton and Gillview interchanges

Interchanges
Number	24
System Interchanges	7
Access Interchanges	17

Structures
Number	82
bridge crossings (culverts excluded)	
At interchanges	32
On cross roads	30
On railways	5
For pedestrians	6
Over rivers	8
For pipe	1

Emergency Services
69 SOS telephones linked to 1 control centre and serving eight local authorities

Cost
R250 million (contract price) for a total of 16 contracts

Population
Approximately 0,8 million within the Ring Road. (See Appendix A for area map).

STANDARD

Cross section
Varies between 4 and 8 lanes

Length
21,2 km : six-lane Northern bypass (Glendower to Sandton)
21,0 km : four-lane Western bypass (Sandton to Maraisburg)
29,8 km : six-lane Southern bypass (Maraisburg to Geldenhuis)
7,6 km : eight-lane Eastern bypass (Geldenhuis to Glendower)
TOTAL = 79,6 km

Interchanges   For interchange spacing, please see the table overpage.

Traffic   For seven-day average daily traffic flow details please see Appendix B.

JUNCTION NUMBER	INTERCHANGE	DISTANCE MILES	KILOMETRES
1 - 2	Buccleuch	2,3	3,7
2 - 3	Sandton	3,0	4,9
3 - 4	Witkoppen	3,5	5,8
4 - 5	Velskoen	3,5	5,6
5 - 6	Boschkop	3,0	4,9
6 - 7	Fairland	1,4	2,3
7 - 8	Gordon Road	1,5	2,4
8 - 9	Maraisburg	3,3	5,3
9 - 10	Soweto	1,4	2,2
10 - 11	Marthinus Smuts	0,8	1,2
11 - 12	Diepkloof	1,5	2,4
12 - 13	Ridgeway	1,6	2,5
13 - 14	Winchester Hills	1,7	2,8
14 - 15	Gillview	1,9	3,1
15 - 16	Oakdene	2,2	3,5
16 - 17	Reading	0,8	1,2
17 - 18	Voortrekker	1,6	2,5
18 - 19	Elands	1,9	3,1
19 - 20	Geldenhuis	1,4	2,2
20 - 21	Van Buuren	1,9	3,0
21 - 22	Gillooly's	1,5	2,4
22 - 23	Glendower	1,9	3,1
23 - 24	Modderfontein	3,1	5,0
24 - 1	Kelvin	2,8	4,5

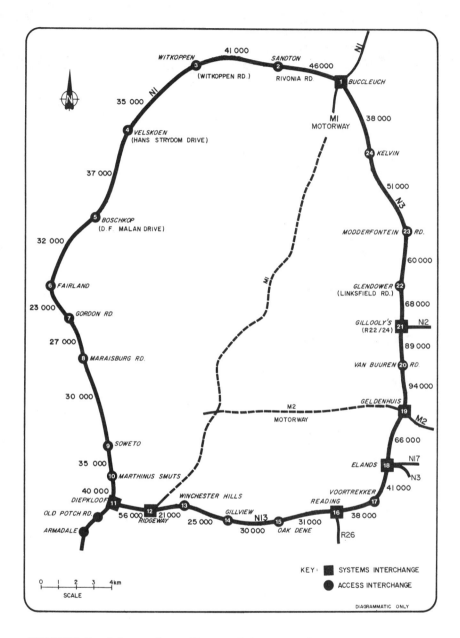

APPENDIX B. Johannesburg National Ring Road –
1987 seven-day average traffic flows

The effect of beltways on urban development: a discussion of US experience

G. T. LATHROP, BCE, MUP, PhD, AICP, MASCE, MITE, Assistant Director of Transportation, City of Charlotte, and K. E. COOK, AB, MA, MAEA, Transportation Research Board, National Research Council

SYNOPSIS. In a context of population growth, market-dominated land development and consumer demand for a "suburban" life-style, orbital motorways can lead to use and development outcomes entirely different from those anticipated. Decentralized development and reduced central area vitality are accompanied by congested motorways and general suburban congestion.

INTRODUCTION

1. The relationship between orbital motorways - or beltways, as they are called in the U.S. - and planning and development is different, we believe, in the U.S. than in many other countries.

2. Anthony Downs, a prominent urbanologist at the Brookings Institution in Washington, has written recently: (Ref. 1):
"For the past few decades, one major vision about how U.S. metropolitan areas ought to be developed has become totally dominant. . . . . This dominant ideal vision is built upon four pillars. Each is a key desire or aspiration shared by nearly all American Households: The first pillar is ownership of detached, single-family homes on spacious lots. Repeated polls show that over 90 percent of all American households would like to own their own homes, and the vast majority want single-family detached units. . . . . . ."

3. "The second pillar is ownership and use of a personal, private automotive vehicle. Every American wants to be able to leap into his or her own car and zoom off on an uncongested road, to wherever he or she wants to go, in total privacy and great comfort - and to arrive there in not more than 20 minutes. . . . . During the five years 1983-1987, the United States added 9.2 million persons to its total human population, but more than twice as many cars and trucks - 20.1 million - to its total number of vehicles in use."

4. "The third pillar of the dominant ideal vision involves the structure of suburban workplaces. They are visualized as consisting predominantly of low-rise office or industrial buildings or shopping centers, in attractively landscaped, park-like settings. Each such structure ought to be surrounded by a large supply of its own free parking. . . . ."

5. "The fourth pillar for this ideal vision concerns governance. Most Americans want to live in small communities with strong local self-governments. They want those governments to control land use, public schools, and other key elements affecting what they perceive as the quality of neighborhood life. . . . ."

6. As is explained below, beltways have been seen as providing relief to the spread of urban congestion which logically follows the realization of this "dominant ideal vision".

7. But metropolitan beltways are not the solutions to congestion which they were anticipated to be. Several unforeseen outcomes have occurred. Beltways around metropolitan areas have become congested, some intolerably. Development has erupted in and near beltways, particularly at interchanges. Beltways have attracted development which might otherwise have occurred in the central part of metropolitan areas.

8. Several phenomena define the context for these outcomes.

9. The population of the United States continues to grow, although the rate of growth is declining (Table 1).

10. That population is becoming more and more urban (Table 1). Importantly, although Table 1 masks the change, many rural residents now are what the U.S. Census terms "rural-nonfarm": they live in rural areas, but they work in town, or small cities, and they are not farmers.

11. Metropolitan areas ("cities" defined by relationships, not by political boundaries) are becoming less dense in both population and employment terms (Table 2). Some are not growing in numeric terms, but they are spreading physically.

TABLE 1

URBAN & RURAL POPULATION GROWTH (Ref. 2)

	URBAN	%	RURAL	%	TOTAL

UNITED STATES

	URBAN	%	RURAL	%	TOTAL
1950	96,847	64.0	54,479	36.0	151,326
1960	125,269	69.9	55,054	30.1	179,323
1970	149,647	73.6	53,565	26.4	203,212
1980	167,051	73.7	59,495	26.3	226,546
1990	190,379	76.2	59,500	23.8	249,879
2000	208,698	77.9	59,051	22.1	267,749
2010	223,469	79.2	58,588	20.8	282,057
$\frac{2010}{1950}$	2.31		1.08		1.86

SOUTH ATLANTIC REGION

	URBAN	%	RURAL	%	TOTAL
1950	10,391	49.1	10,791	50.9	21,182
2010	42,320	76.8	12,791	23.2	55,111
$\frac{2010}{1950}$	4.07		1.19		2.60

TABLE 2

POPULATION AND EMPLOYMENT DENSITIES OF CENTRAL CITIES AND URBAN FRINGES: 25 LARGEST URBANIZED AREAS (Ref. 2)

		Central City	Urban Fringe	UrbArea
Residents	1970	8215	3092	4298
per Sq.Mi.	1980	6859	2697	3548
Percent Change				
Density		-16.5	-12.8	-17.4
Number		- 7.1	+16.1	+ 5.7
Workers	1970	3751	1013	1658
per Sq.Mi.	1980	3690	1047	1587
Percent Change				
Density		- 1.6	+ 3.3	- 4.2
Number		+ 9.4	+37.6	+22.6

145

12. The number of automobiles per person and per household continues to increase, a larger percentage of adults is in the work force, and dependence on the privately owned automobile continues to increase (Table 3). Radial commuting patterns, which transit traditionally has served well, are shrinking relative to circumferential commuting in the lower density areas. (Table 4 and 5)

13. While beltways have been a catalyst for centripetal relocation of growth and circumferential commuting, both have occurred and are occurring, throughout metropolitan areas in the U.S., regardless of the presence of a beltway.

THE DEVELOPMENT OF BELTWAYS IN THE UNITED STATES.

14. Prior to World War II, highways were built to interconnect towns and urban areas and to connect rural areas to towns with a network of all weather roads. This period is often referred to as "getting the nation out of the mud." As the number of vehicles increased after the war period, urban traffic volumes increased. There was an increasing proportion of trucks. Traffic turning movements, cross traffic and access to and egress from commercial establishments increased delay to such an extent that the need to separate through traffic from local traffic became apparent.

15. The congestion, in central areas, led to the concept of highway by-passes around the towns and cities.

16. Though strongly endorsed by highway users, by-passes frequently were resisted by local merchants who feared loss of business. To justify their construction, economic impact studies of by-passes were performed. They considered such issues as: people and businesses displaced by right-of-way takings, proximity effects of by-passes, effects on community cohesion, changes in accessibility, effects on the transportation disadvantaged, and, later, environmental effects. The studies indicated that highway oriented businesses that were by-passed were adversely effected, but that commercial activity, especially highway oriented business, relocated to the new facility. Manufacturing also was quick to relocate along the by-passes. Managers of such plants indicated that two major considerations were the ease of getting raw materials and finished goods to and from the plant and improved accessibility for employees.

TABLE 3

DISTRIBUTION OF PERSON TRIPS BY MODE OF TRANSPORTATION AND PURPOSE (PERCENTAGE) (Ref. 3)

	1977 Auto	Public Transit	1983 Auto	Public Transit
Earning a Living	89.5	5.2	87.1	4.5
Family & Personal Business	89.2	1.6	87.9	1.1
Civic, Educ. & Relig.	57.9	4.8	55.9	4.7
Soc. and Recreat.	87.6	1.5	81.2	1.6
Other	80.3	7.5	83.7	0.8
Total	84.7	3.0	82.0	2.5

(Percentages do not total to 100 because of omitted minor modes)

TABLE 4

PLACE OF WORK BY PLACE OF RESIDENCE, U.S. URBANIZED AREAS (PERCENTAGES) (Ref. 2)

Live	Work	1970	1980	Change
Central City	CBD	6.3	5.5	-0.8
	Other Cent	36.5	31.7	-4.8
	Fringe	10.5	8.6	-1.9
	Sub-Total	53.3	45.8	-7.5
Fringe	CBD	2.6	3.3	+0.7
	Other Cent	13.2	15.3	+2.1
	Fringe	30.9	35.6	+4.7
	Sub-Total	46.7	54.2	+7.5
	TOTAL	100	100	

TABLE 5

USE OF TRANSIT TO WORK IN VARIOUS MARKETS IN
U.S. METROPOLITAN AREAS (Ref. 3)

Live	Work	Percent 1970	Percent 1980
Central	CBD	46.3	33.4
City (CC)	CC/not CBD	17.3	13.1
	Outside CC	9.8	5.5
Suburb	CBD	32.8	32.1
	CC/not CBD	7.6	7.1
	Outside CC	4.4	2.1
TOTAL		13.4	8.7
(N*1000)		(5,694)	(4,869)

17. Beltways generally were built on relatively undeveloped land that displaced comparatively few businesses and residences. Increasingly they were seen as the only feasible alternative when the "freeway revolt" occurred in reaction to the massive dislocations caused by trying to "punch interstate highways through residential communities." The continuing urbanization of the population caused other changes in the cities. Commercial and shopping districts emerged around the new by-passes because of the ease of access. By-passes that were constructed to circumvent the congestion of the downtown core quickly became the main thoroughfares for new commercial development.

18. Full beltways were a logical extension of the bypass. The ring and radial road pattern developed as the geometric reflection of declining density around an urban core. In some cities, such as Austin, Texas, primary highway routes outside the Interstate beltway have been reconstructed to act as outer beltways. In Boston, when Interstate Route 95 (old Primary Route 128) became congested, because of development nearby, a new outer beltway, Interstate 495, was constructed. In 1979, 80 percent of office space in Boston was located in the center city. By 1984, office space was equally divided between City center and the suburbs, and since that time, suburban office space is rapidly increasing compared to downtown. Recent planning studies indicate that for every square foot of new office space constructed in downtown Boston, there are ten square feet constructed along or outside Route 128.

BELTWAYS AND DEVELOPMENT

Axioms:

19. Increasing the accessibility to land which has potential for economic development, through higher capacity transportation systems, will permit and encourage increased intensity of land use and increased travel, compromising the effect of added capacity in reducing congestion.

20. Transportation is essentially a two dimensional system (even though it may be at ground level, above ground or below ground); urban land development is a three dimensional system. As transportation service capacity increases, urban development increases vertically.

21. Where the transportation system collects users from low density areas and delivers them to high density areas, travel will be more concentrated at the high density area and congestion will result, to the limits of capacity.

22. If any system offers higher service characteristics than existing systems and has excess capacity, it will divert traffic from the lower service system to the higher service system. Drivers are willing to travel a more circuitous route on freeways to save time. Traffic will divert to freeways up to the point that the expected travel time on the freeway approaches that on arterials.

23. More remote land, served by beltways or bypasses, gains accessibility and is developed due to its lower cost in comparison to land closer to the center. Combined with population growth and the desire for a suburban home, this increased accessibility propelled suburban growth. Beltways became self-justifying as they served local suburban as well as through traffic.

24. Because of lower development costs and a growing population (labor supply) and market in the suburbs, industry and business began to locate in the suburbs and to congregate into activity centers. Around the urban core, older housing became less attractive than new housing in the suburbs, and rural immigrants and minority groups, who had less income, found residences in the older inner city housing. This socioeconomic population shift has continued over several decades. With affluence and a market for labor in the suburbs, the rate of growth of commercial development in the suburbs is accelerating at several times the rate in the urban core. Urban cores are losing population (the city of St. Louis lost 35 percent of its

population in the last ten years, while the suburbs continued to grow).

25. In the past, there has been political support for reinvestment in declining central business districts and the use of transportation and other publicly funded construction programs to stem the flight of business from the central business districts to the suburbs. The transportation emphasis in urban areas continued to be improving access to central business districts by improving transportation systems in a radiating spoke fashion to feed travel from the suburbs to the central core and by building and subsidizing public transit systems. To support the radial transportation system, urban planning was directed toward developing high density corridors with low density wedges between them. This pattern emphasized developing land along the transportation corridor, for both business and residences, leaving land wedges between these corridors at lower density. This concept has not been consistent with land market behavior and developers' or consumers' desires, and development in the suburbs continues to sprawl.

26. Labor distribution is not uniform due to the differences in affordable housing. This exacerbates journey to work commuting. The affluent have a wide variety of housing options throughout the metropolitan area. However, the location of their job opportunities may be limited, often to the central core. This is especially true for professionals in activities such as finance, insurance, law, planning and marketing.

27. Central core cities, with investments from the federal, state and private sectors, continue to grow through urban renewal and replacement of older manufacturing and office space by new higher density office space. Although land and development costs are higher than those in the suburbs, these are offset to a significant extent by the availability of existing infrastructure, the high accessibility provided by the existing transportation system and various public inducements. The market for space in the urban core also may be perceived as more stable because of its size and the flexibility available through marginal price adjustments. Suburban space faces a more "all-or-nothing" environment.

28. While the primary transportation linkages have been to the urban core, rapid growth in suburban activity centers has resulted in a very rapid increase in cross suburban, intra-regional travel. In many cities center-oriented trips now account for less than 20 percent of all travel. Suburb to suburb commuter trips now often exceed 50 percent of travel, and, combined with non-work

travel, all suburb to suburb travel accounts for as much as 2/3rds of total travel.

Beltway Operating Characteristics

29. Beltways and freeways, in order to operate efficiently, must be supported by a system of arterial highways. Lacking such supporting systems, the beltway or freeway will function as an arterial for short trips, and traffic congestion will result because the system is performing both the through movement and the local collection and distribution functions.

30. When freeways and beltways exceed their design capacity (e.g. 2,000 vehicles per lane per hour), the total throughput of traffic rapidly and absolutely decreases.

31. Traffic entry and exit at interchanges on beltways and freeways makes them the primary points of traffic congestion. In order to control traffic flow, it is necessary to control the interference and access points by: (1) limiting the number of access points, (2) limiting the flow of traffic through the access points with some form of ramp metering, and/or (3) designing access ramps and directional interchanges that will minimize traffic friction and weaving movements.

32. After a system is in place, and the design parameters for access to the system have been established, it is difficult to close access points, or to prevent new access points from being created as long as they conform to established standards. To close or meter access points is seen as inequitable; persons entering the system upstream have full access, while those entering downstream, when the system has become congested, are limited in their ability to use the system. While generally it has not been successful to take an existing lane and convert it into a high occupancy vehicle lane (e.g. diamond lanes in Los Angeles), the construction of new reserved high occupancy vehicle lanes has been successful and acceptable. (If the bus and car pool traffic on such reserved lanes does not meet the public expectation in volume, there will be continual pressure to reduce the number of passengers per vehicle required to use the reserved lanes.)

33. Many of the traffic congestion points, in urban areas, now are outside the beltways and are due, to a great extent, to the conflict between commuters destined for the CBD and those destined for other suburban locations along the beltways. Peak travel periods, in the morning and evening, start earlier and last longer.

34. The availability of high speed and high capacity
service and the failure to charge individual vehicles the
full costs of such service, has induced an expansion of
truck traffic, on freeways and beltways, many times larger
than anticipated. For example, 20-year forecasts, on most
interstate routes in or between major urban areas, were
estimated at 8-15 percent truck traffic for freeway
design. Many routes are currently experiencing 40-60
percent truck traffic.

Beltway Congestion

35. Most urban traffic congestion problems will have to
be ameliorated through the creation of new transportation
facilities and services. Transportation systems management
- attempts to control the demand for and supply of
transportation services on existing facilities - will
address only a small percentage of congestion. A solution
will be achieved only through new investment in
transportation facilities and services and pricing of those
facilities and services, to the users and developers, at
their true costs.

36. Traditionally, transportation has seen itself as
serving land development. Transportation investment, where
there is no viable demand for economic development, does
not induce development (deliberate transportation
investments in the Appalachian Mountain region did not
produce any significant long-term economic development).
Development opportunities are reflected in speculative land
prices, and attempts to acquire developable land, as part
of proposed new transportation corridors, in order to sell
or develop later to help cover the costs of the
transportation improvement, have not proven successful
(value capture has not worked).

37. Traditionally, in the U.S., transportation services
have been priced to the users who are the direct
beneficiaries of the services. Local governments have
tried to control land development through zoning and other
use controls. It has become increasingly difficult to
respond to problems of urban congestion through the
publicly funded, user charge approach to improving
transportation services, and land use control has been
ineffective in controlling the development of land or
balancing it with infrastructure needs and the public's
perception of their desired quality of life. The question
now is whether we might better approach the problem by
recognizing that the free market is the best arbitrator of
the development of land, but that the "free" market should
reflect the true costs of development options. Both
transportation users and developers, who create the need

for transportation, if charged the true transportation related costs, might change the way publicly funded transportation facilities and services are allocated.

38. A major problem in attempting to develop a transportation system in an urban area, or to establish consistent land use controls throughout a metropolitan area, is the multiplicity of governments and concomitant diffusion of power. Different local governmental units have different objectives and economic and social bases, and it is difficult to establish transportation systems or control land development in a unified manner. Furthermore, the approach to planning in the United States has been reactive rather than proactive. Eighty-five percent of capital investments are made by the private sector and only about 15 percent by government. The main driving forces are determined by the market place. Developers of land react to the market, and governmental services are reactive to development. This means that urban form and transportation systems follow demand and that there is little public control or influence. Accordingly the role of government planning, in a democratic, market driven economy, is primarily negative, putting constraints on development.

BELTWAYS AND FUTURE LAND DEVELOPMENT

39. In 1991, the U.S. Congress will be faced with the issue of what to do with the Federal Highway Trust Fund that has been used during the past thirty years to pay for the construction of the interstate highway system and the reconstruction of the heavily traveled federal aided primary highway routes. These roadways constitute an estimated 170,000 mile network of high volume, nationally significant highways. Federal urban public transportation programs also are funded with a combination of one cent from the Highway Trust Fund and general revenue funds. Therefore, the future of federally funded urban public transportation also is at issue.

40. There has been a substantial disenchantment by the public in the ability of the federal government to deliver services in an effective and efficient manner. The federal deficit continues to rise, Congress remains indecisive, and the executive branch continues the policy of reducing the federal role in government.

41. The states, in turn, have increased their taxation and responsibility for transportation and other governmental services. Their resistance to donor state status in terms of federal taxes also is increasing. They

want the taxes that they contribute returned to them in like proportion.

42. In such an environment, it is difficult to anticipate that the federal government will take a leadership role in helping to solve urban transportation congestion problems. The best that can be hoped for is that the federal trust fund will be preserved and that it will be used for the continued rehabilitation of a national network of interstate and key primary highways. The states and local governments will have to resolve the problems of suburban traffic congestion and the further decline of the city center. To accomplish this, they will have to trade some resources with rural interests that are still strong in state legislatures.

43. As a result, there will be renewed interest in building bypasses around rural communities and in upgrading the quality of highways in rural areas (under the pretext of economic development). In exchange, urban areas will receive new beltways and cross suburb connectors.

44. We do not foresee any dramatic change in the current patterns of low density housing in the suburbs or the increase of business and commercial activity in centers in the suburbs. Central business district oriented traffic, in most urban areas, will remain relatively constant in the future. Air quality and environmental considerations and the lack of financial resources needed to preserve and improve the decaying city core infrastructure will prevent the central business districts from further expanding.

45. Concern will increase about heavy truck traffic compounding commuter traffic congestion. Combined with safety considerations, there will be renewed interest in separating through truck traffic from local automobile traffic.

46. Land development will continue to leap frog to under-developed land in the far suburbs and will be followed by gradual "build out" from the center. Rather than trying to prevent or control this pattern, the counties and cities will do well to require reservation and dedication of land for transportation corridors.

47. In suburban areas, where "build out" is occurring, exaction fees will be assessed and developers will be required to pay for increasing transportation systems management (both transportation demand and supply). In addition, developers will be responsible for an increasing

portion of the infrastructure costs that their development requires, both on-site and at off-site congestion points.

48. One of the greatest problems, to be faced with suburbanization and its lower densities, is the cost of providing needed public services. Costs increase geometrically, not arithmetically, with the result that home ownership may be put beyond the financial resources of an increasing proportion of the population.

49. In forecasting the future, we tend to err in believing that trends will continue and that we should plan to meet such exigencies. We built the interstate highway system, but did not foresee how it would induce travel. We built beltways, but did not see how they would effect settlement patterns. We built tall buildings in cities and did not foresee what they would do to the quality of life around them. We have come to realize that with each new development in human settlement patterns, there also come disadvantages.

50. In the future, we will approach change by incremental tinkering rather than grandiose future visions. From what we see now, our cities are developing in a wave pattern like those from a rock dropped into a still pool of water. At first the concentric waves are abrupt but quickly they moderate and finally die out. Perhaps the future of our urban environments will follow a similar pattern. Right now, the traffic pressure is sufficient to warrant the construction of at least another outer beltway around most of our congested, suburbanizing cities. However, these new beltways will be built in incremental, self justified segments rather than as a total system. Beltways provide accessibility, and accessibility induces development in an urban environment.

SUMMARY

51. We have attempted, in a very brief way, to describe the relationship between beltways-orbital motorways - and the development of land near urban areas in the United States.

52. In conjunction with the four "pillars" described by Downs, continued growth of and shift in population, and the dominance of market forces in land development, are leading to a decentralization of urban activity which is facilitated by the relative accessibility provided by beltways. Failure to provide transportation capacity, in addition to beltways, has led to congestion of the entire suburban transport system.

53. It is not possible for us to generalize about the possibility or probability of similar outcomes outside of the United States.

54. If we have been reasonably successful in identifying the causes of and factors contributing to extensive development and suburban, and beltway, congestion, then others can judge the extent to which they occur in their own countries and contexts, and draw conclusions about the extent to which similar outcomes will occur.

55. Our sense is that policies, control, governance and economics are different than those in the United States. But, if the descriptions of U.S. values and vision "ring a bell", and development controls are drifting toward the laissez faire, market-dominated situation we have described, then beltways - and suburban highway accessibility - can have a profound effect on the shape of urban development.

REFERENCES

1. DOWNS, ANTHONY, "The Need For a New Vision for the Development of Large U.S. Metropolitan Areas", The Brookings Institution, Washington, D.C., 1989.
2. LOWRY, IRA S., "Planning for Urban Sprawl", A Look Ahead: Year 2020, Special Report 220, Transportation Research Board, National Research Council, Washington, D.C., 1988.
3. RENO, ARLEE T., "Personal Mobility in the United States", A Look Ahead: Year 2020, Special Report 220, Transportation Research Board, National Research Council, Washington, D.C., 1988.

Discussion

A.J.E. ROSE, Mott MacDonald Civil Ltd, Winchester
It is essential that we make the best use of the
existing highway network, and there are many ways in
which this can be addressed. Considering the
removal of interchanges on the Charlotte Beltway or
the introduction of additional interchanges in the
Johannesburg beltway together with management
systems are only short term palliatives. The longer
term solution seems to be the acceptance of urban
expansion and the inevitable need for a second outer
orbital in part or in full. The questions remaining
are not 'what?' but 'how?' and 'when?'.

R.D. RIDDETT, Travers Morgan, London
Very high traffic volumes on outer orbital roads
mean that we should take a look at public transport
to cater for some of the demand. Perhaps all new
construction improving existing orbital roads and
motorways should consist of high occupancy vehicle
lanes. Also, although trips using these roads are
basically radial in nature and highly variable in
pattern, a large number of identical trips take
place every day. It seems possible that some of them
could be handled by other means than one person cars.

B. PARKER, G. Brian Parker, Twickenham
I am surprised by Dr Lathrop's strong view that
roads do not induce development, but I do not
disagree. I trust that those who disagree will
speak up, because this is an important point about
which we should argue here.
    I also wish to comment on Dr Lathrop's assertion
that vehicle ownership rather than beltways induces
trip making. Would he agree that beltways tend to
stimulate vehicle ownership, and thereby indirectly
induce trip making?
    Dr Lathrop mentioned that the dispersal of trip
origins and destinations makes it difficult to serve
movements of high occupancy vehicles. Does he agree

that beltways tend to encourage dispersal of trips,
rendering the problem more difficult to solve?

S.N. MUSTOW, Consultant, Sutton Coldfield

1.   While it may be true that the provision of a
     road will not cause development, it is
     nevertheless true that local government and
     other agencies will see the road, particularly
     the areas near interchanges, as offering
     development opportunities.

2.   Reference has been made to congestion on
     access roads to beltways (which will be
     greater with fewer interchanges). What is
     your experience in regard to adequate
     co-ordinated planning of access road
     improvement with beltway development?

PROFESSOR R.J. BROWN, Transportation Planning Associates, Birmingham

It was Wilfred Owen who said that 'transport is a
necessary but not sufficient element in the
development process'. It is in fact a sine qua non
- transport has to be considered together with
capital, labour, finance, power, etc. before the
development process takes off. Without the proper
level of investment in the transport infrastructure
we are unlikely to release the forces that stimulate
the economic development process.

J. WOOTTON, Wootton Jeffreys Consultants Ltd, Woking

At the end of the presentation of my paper I
suggested three courses of action. One of them
concerned the point that has been discussed again,
namely the co-ordination of land use and transport
activities. The action I suggested was to find some
way of co-ordinating investment in both development
and transport. The point made by Mr Lathrop that in
the USA the public sector only invests 15% of the
infrastructure costs seems to add weight to my
point, as it implies that most of the investment in
the built environment comes from private sources.
Thus, co-ordination of investment in development and
transport is likely to be difficult when there are
so many independent sources of finance. An
appropriate mechanism might help, so I suggest, for
debate, the concept of a development bond.

Each piece of land would have a development bond
which is held by the person who owns the land. The
development bond must have a number of other
characteristics. The first is that it has a value,
and that value is established on the basis of
change, i.e. change from one type of land activity
to another or change in accessibility. Clearly, the
value has to be determined independently: this might

be done through an independent body, local authority or similar organisation. The second characteristic is that the value is only realised when a development takes place, whether it is a road, a change in land activity, a change of intensity, etc. This would result in an automatic payment to the local authority if the value of the bond increases due to the development or by the local authority to the holder of the bond if it decreases. The amounts collected by the local authority would be held on account and used only for infrastructure improvements.

M. SIMMONS, London Planning Advisory Committee, Romford

Regarding relationships between transport infrastructure and development: we tend to generalize; the orbital point is the nature of the regional economy. The situation is quite different as between peripheral or remote regions where new infrastructure has a disappointing effect in stimulating growth, and metropolitan or 'core' regions where the scale and locations of development is closely related to the extent to which new infrastructure provides the accessability to sustain growth. In such regions planning of transportation needs to be an essential component of development policy, related to the forces of growth, if it is to be accommodated in a successful way so that the economy of that region remains competitive, the example being London.

R.A. STANWAY

Because the Johannesburg National Ring Road is located far out from the CBD, we tend to find a far more dispersed origin-destination pattern which does not lend itself to public transport movements. Nevertheless, many Combi-Taxi or paratransit vehicles do make use of the facility fairly effectively without the need for high occupancy vehicle lanes.

The experience with development location associated with the Johannesburg National Ring Road has been that residential concentrations as well as commercial nodes have tended to spring up close to access interchanges. It is also clear that nodes that existed prior to the opening of the Ring Road have developed faster and more successfully than those without good access to the Ring Road.

Our experience with respect to access road improvement associated with Ring Road congestion has been good. Interchange spacings are large and ramp volumes at the diamond access interchanges are higher than it is practical to handle with ramp metering. Nevertheless, thorough monitoring, short-term improvements and 'turning' at these interchanges have so far been successful in balancing volumes in the Ring Road with access road volumes. In the longer term one might have to consider extending the possible Freeway Traffic Management concepts far back along the access road corridors.

DR G.T. LATHROP

Certainly Professor Brown and Mr Owen are correct. The emphasis to which I take exception is that which asserts that 'transportation causes development', or that an investment in transportation, such as that represented by the orbital motorway, will <u>cause</u> development. It might be more accurate to assert that the motorway influenced the <u>location</u> of development, in a context in which the development was imminent.

In response to Mr Riddett, our experience in the USA indicates that public transportation and ridesharing do not perform well in suburban contexts (such as that were the orbitals typically are located) because of the diffuse pattern of trip origins and destinations. The pattern is 'many to many' not 'many to one', as with radial movements, much less 'one to one'.

In response to Mr Parker, I would not argue with the assertion that beltways tend to stimulate vehicle ownership but the influence is weak and indirect. I would argue that a number of other factors are stronger in the decision to own. (See Lerman and Ben-Akiva, 'Disaggregate behavioral model of automobile ownership', <u>Transportation Research Record</u> 569, 1976, and a related paper by Lerman in <u>Transportation Research Record</u> 610.) As development occurs and the opportunities for activities become more diffuse, a private automobile, and the mobility it affords, become more desirable. Beltways automobile more desirable, and so on.

In reply to Mr Simmons: as a generality, I agree. But, again, I would put transport well down in any hierarchy of causes. I certainly cannot refer to London with any authority, but I recall reading that the public transport system <u>and</u> the street system in the central areas are <u>badly</u> overburdened: but doesn't development continue?

Summing up

J. A. L. DAWSON, Director of Roads, Scottish Development
Department

SYNOPSIS. The Rapporteur has summed up the
main themes discussed at Conference and the major
propositions.

INTRODUCTION

1. It is not the role of the Rapporteur to sum up
each of the papers or the workshop discussions but
to bring out the main points and themes that have
arisen in this Orbital Motorways Conference. It
was the late Alan Brant who conceived this
Conference and the Conference has paid frequent
tribute to him. In formulating the Conference, his
hypothesis was that the experience in Britain with
the M25 could not be an isolated event in
international terms. He was clearly right. Two of
the case studies presented to the Conference show
flows and congestion on orbital roads higher than
that seen on the M25 and major forward investment
plans in these countries are in hand to address the
problems.

OBJECTIVES

2. In terms of customer demand, major orbital
roads are clearly a success. But the Conference
discussed particularly the objectives against which
the success or failure or orbital roads should be
considered. Unlike many conferences on roads,
there has been more than just an exchange of
valuable new ideas and techniques. There has been
much use of the knowledge and different experience
of others in an attempt to understand the transport
and land use dynamics at work in metropolitan areas.

3.   Professor Hall set the stage in his keynote
address.   He showed the ring road concept was so
successful it had existed for centuries.   In his
remarkable history he showed how the development
of motorways and orbital roads ping ponged between
the old world and the new until the controlled
access, grade separated orbital motorway was born
in the first half of the 20th century.   Many of
those attending the Conference from Britain found
his reminder of the birth of the British motorway in
the second half of the century timely in the light of
the current programme expansion.

4.   Professor Hall first asked the question "what
are orbital motorways for?"   The general answer
given - using Mr Simmons' words in his presentation
on the M25 case history - was that they had been
"narrowly conceived" as bypass roads.   Indeed, the
Conference has seen in the various case studies how
all the orbital roads began as unconnected bypasses
which were joined together.

THE IMPACT OF THE ORBITAL ROAD ON LAND USE

5.   That does not mean the roads were not being
sited to serve existing or forecast land use.
M. Koenig showed how the orbital road served new
growth points in Paris.   In London, Professor Hall
showed how Abercromby had thought in terms not
just of traffic relief but also in terms of "revealing
latent organic structure" and "defining urban
villages".   Memorably, he said the M25 was " the
route Abercromby would have taken if he had had
time to think about it".

6.   On the first day of the Conference, all four
speakers used phrases on the lines of "no-one had
much thought about the impact of roads on
development" - a sentiment Mr Williams speaking of
Washington also shared on the second day.   The
development attraction of new orbital roads, "like
iron filings to a magnet", was shown in many slides.
But it is a long step from showing some localised
major developments alongside some motorway
junctions in continually growing economies to arguing
that the city is being reshaped and pulled outwards.
Speakers nonetheless addressed this more difficult
question.   In particular, Professor Hall quoted
comparative US work on cities with and without
beltways and the conclusion reached of small but
statistically significant effects arising from the
orbital road.

162

7. Mr Wooton argued that the highly localised siting of development right adjacent to the motorway induces significant extra traffic volume on the motorway. This type of traffic generation, where the motorway is the development's access road, is a different kind of generation from general traffic growth in the vicinity of the motorway. Mr Williams showed how one of the first beltways had turned from a bypass into "Washington's Main Street". Mr Leux showed how Johannesburg's orbital road, without quality radial links into the central business district, had added to the development of the peripheral regional business districts and decentralisation generally.

8. The Conference's regional planning workshop in particular discussed these land use and public transport issues with much reference to Paris and London.

THE TRAFFIC FUNCTION OF ORBITAL ROADS

9. If no one had thought much about induced development, what about the traffic function of these roads? The position and number of orbitals clearly matters: the function of the peripherique with a radius of 10km does not serve the same market as the M25 with a radius of 30km. We were reminded from the floor of TRRLs' 1960's theoretical work on the efficiency of various network configurations and the separate traffic functions of internal distribution, regional distribution and the national bypass.

10. In the real world, a very similar story was told in all the case studies. High flows and congestion on the orbital motorways were occurring because the national and regional distribution functions had in practice got fouled up with the internal distribution of local traffic. Despite the lofty statements about the higher level functions, there were in practice too many junctions on the orbitals attracting too much local traffic to serve these higher level functions properly. In essence, "local considerations dominated the outcome".

BUILDABILITY

11. The confusion of functions perhaps would not have mattered if original road investment plans had been maintained. In the route selection workshop,

there was much talk about "buildability". In
Washington, as in London and Paris, the original
plans became amended or halted in the face of
opposition. In London, the M25 and North Circular
Road represent the art of the buildable. Increasing
use of cut and cover tunnelling is necessary to gain
public acceptance. In Paris, cut and cover is giving
way to bored tunnel as necessary to protect some
sensitive environments and carry on building. The
existing right of way in Washington established by
the beltway is a practical fact of life for the future
in considering what further improvements may be
buildable. In Johannesburg, long term major
developments seem likely to be highly controversial
with "who funds?" issues also disputed between
Agencies.

OPERATIONAL CONTROL

12. The need for improved operational control was
undisputed. Mr Wooton's presentation gave us a
clear exposition of why. The techniques can
optimise capacity and bring higher levels of safety at
very high rates of return with few environmental
problems. He illustrated, for example, how access
control could increase flows by up to 5% and cut
journey times by 20%. He quoted Dutch experience
that automatic systems give benefits 4 times higher
than static systems and that drivers respond well to
the information that is relevant. He pointed to the
new British "Autoguide" legislation and systems
being developed in Berlin, London and Lyons and
to the availability of the new European communication
networks.

13. In the operational control workshop, there was
much debate on institutions. Very real
organisational and procurement problems emerged.
Specifying and procuring systems in a fast moving
arena, and deciding who is going to operate them, is
at least as challenging as the technology of the
systems themselves. Assistant Chief
Constable Mannion from Strathclyde gave the
operational control workshop a practical diagnosis of
the problem, pointing out that 75% of the 17,000 calls
a year on the Glasgow CITRAC system are
breakdowns. Of those, about 30% result in CITRAC
operation. Nationally that probably implies around
one million breakdowns per annum on Britain's
motorways. The workshop discussed whether the

scale of delay costs imposed by breakdowns argued for stricter vehicle condition controls or even penalties.

THE FUTURE

14. So what of the future? In the Ile de France M. Koenig said traffic growth had accelerated from 3 to 6% pa and traffic flows were "not expected to stabilise in the immediate future" with the Peripherique already carrying 220,000 vehicles per day on dual-4's. In Washington, John Williams reported flows on the dual-4 beltway of 225,000 forecast to rise to 300,000, 60% of which were internal trips. In London, John Wooton pointed to the UK as having the slowest rate of growth of European countries except Denmark and forecast strong growth in demand. In Johannesburg, some sections of the orbital were carrying 90,000 with the potential for substantial growth given the developing country characteristics of South Africa.

15. In all 4 cases, major programmes were in train or being considered. The Ile de France's basic programme had been increased by about 50%. In addition, there were major programmes on the A84 and la Francilliene. Further there was a private sector project being discussed consisting of a tolled underground orbital and associated radials to take 15% of Paris' traffic. In Washington, mid-range proposals are being considered either to widen to 10 lanes or reduce lanes to 11 feet; the long range view is examining widening to Toronto style 3-4-4-3 or an additional western and eastern bypass of the City. On the M25, there is dual-4 widening and other bypass proposals on the table or being studied. In Johannesburg, there are major institutional questions over funding and function with a recognition of the need to investigate major improvements.

FUNDING AND DEMAND MANAGEMENT

16. All these plans and proposals have major expenditure demands in common and the funding issues, who pays and how, were discussed. The Conference also discussed demand management and priority management with electronic road pricing, high occupancy vehicle lanes, and ramp metering. Traffic calming undertaken in association with schemes was discussed. In particular, M. Koenig identified the "constant traffic principle" that may

allow the underground network in Paris to proceed. In this principle, the new capacity is not additional but replacement capacity. The road space relieved is "re-conquered" for amenity or public transport use. John Wooton gave a colourful picture of how the information technology may transform transport over the next 30 years.

CONCLUSION

17. The Conference generated a number of main propositions:

> 1. Radical restraint actions are possible but the world is set on a path for significant traffic growth and yet faster congestion growth in the developed Metropolitan areas.

> 2. Orbital motorways are heavily used, can work, and public transport is unlikely to cope with the unfocused journeys that they serve.

> 3. Public transport and land use planning for future orbital motorway projects needs to be done closer together with the impact on the whole city considered better.

> 4. The capacity for national and regional through traffic needs to be protected.

> 5. Sophisticated operational control is essential for handling large traffic volumes safely and efficiently. Information technology needs to be harnessed more energetically and the organisational framework needs to be better addressed.

> 6. Road pricing, for finance and/or congestion management, will need to be widely considered.

> 7. Communities will look for increasing quality in the environmental treatment of road schemes if they are to be buildable.

In closing I would like to thank all the speakers on the platform, from the floor and in the workshops for making this such a thoughtful Conference.